D1591654

WILDFLOWERS OF INDIANA

Wildflowers OF INDIANA

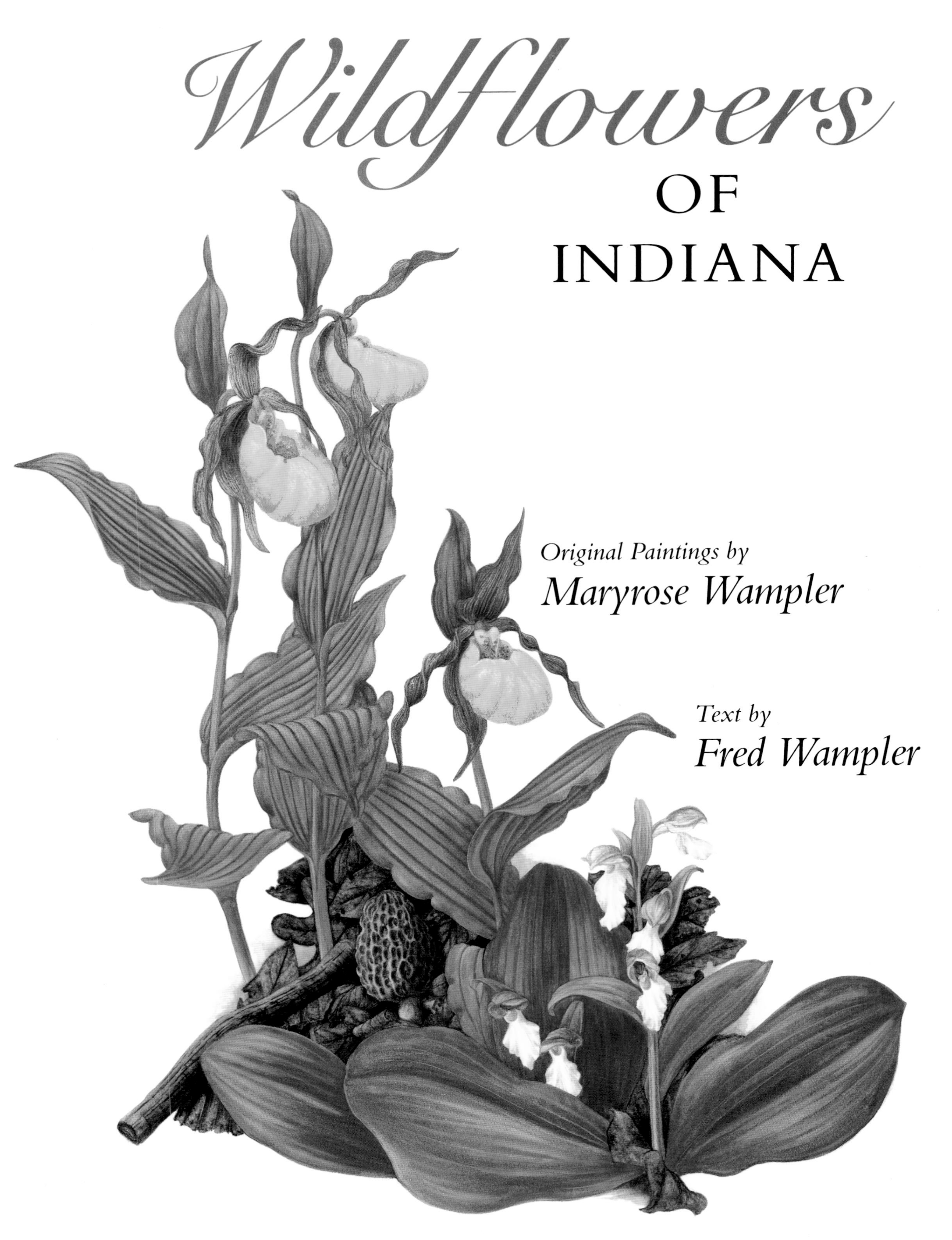

Original Paintings by
Maryrose Wampler

Text by
Fred Wampler

INDIANA UNIVERSITY PRESS • BLOOMINGTON AND INDIANAPOLIS

Manufactured in Japan

Library of Congress Cataloging-in-Publication Data

Wampler, Maryrose.
Wildflowers of Indiana.

Bibliography: p.
Includes index.
1. Wild flowers—Indiana. 2. Wild flowers—Indiana—Pictorial works. 3. Botanical illustration—Indiana. I. Wampler, Fred. II. Title.
QK159.W35 1988 582.13'09772 88-45102
ISBN 0-253-36573-2

1 2 3 4 5 92 91 90 89 88

To Ruth Davison and Skip and Kay Beth Harrell
for their patient support of our artistic endeavors.

CONTENTS

PREFACE

Maryrose began the first painting for *Wildflowers of Indiana* in the fall of 1983. During the ensuing years, our task of locating and obtaining specimens and making accurate identification of species has been greatly facilitated by the many helpful, knowledgeable people we have met in the course of our work. The bits of folklore, information, and expressions of interest have been helpful and encouraging to us, and we should like to specifically name some of our sources of assistance.

Lois Metino Gray, a former naturalist with the Indiana State Parks Service, is responsible for the inclusion of several of the rare plants Maryrose has painted and has served as a consultant in the preparation of the text.

We are indebted to Bill Walters, Director, Division of State Parks, for granting permission to collect specimens in the state parks, and to Mike Homoya, a botanist with the Indiana Department of Natural Resources, for his patience and assistance in dealing with species identification. Personnel in the state parks have been invaluable sources of information about bloom times and locations of plants we wanted to include in our book. The officials at Indiana Dunes National Lakeshore allowed us to photograph rare plants in their delicate natural habitat. Clarence and Esther Hurley extended both hospitality and assistance in Maryrose's work at Smith Cemetery. As a result of the help and interest of these people, this book is essentially an illustration of the wildflowers in the Indiana parks.

We had a lot of fun comparing notes with Kay McCrary, who is currently writing a comprehensive fieldguide of Indiana wildflowers. In addition to making the resources of the Indiana University Herbarium available to us, Kay was generous in supplying information on bloom times and plant locations. Others who have shared their knowledge and resource materials with us are: Henry and Pearl Berry, Kathleen Berry, Keith Craddick, Margaret Dring, Ercell Fritz, Charles Hagen, Sam Miller, Emma Pitcher, Tony Sparks, Sharon Sprague, and Jane Taubensee.

Some of the friends we have made in the course of our work have shared not only their knowledge and interest in wildflowers, but their property as well. We wish especially to thank Bob and Laura Kern for their substantial contribution to our efforts. Their Christmas tree farm in Fulton County supports a rich and varied habitat for flora ranging from tamarack swamps to dry, sandy fields. Avid and knowledgeable wildflower enthusiasts, they were able to locate unusual plants for us and to notify us when they were at peak bloom. They also generously supplied us with modern facilities and a parking place for our camper.

We wish to thank Marion and Mary Young for access to the Young family farm in Brummett's Creek valley in Monroe County and Michael and Darlene Barth for the use of their property on the Whitewater River near Metamora. We found both of these areas to be unusually rich in both variety and quantity of wildflower population. Other generous people who shared wildflowers growing on their property were Norman and Betty Lucas, Virginia Lyon, Howard Sanders, and John and Mary Ann Seeber.

Some of our friends even lifted specimens and brought them to us. We owe the inclusion of *Spiranthes tuberosa* to Roger Beckman, who found this rare jewel growing in his backyard. The elusive Scarlet Pimpernel was contributed by Loren and Gertie Shahin, whose interest in our book from its inception has been a great encouragement to us.

Most of my efforts at writing the text for this book were conducted in a supine position with snacks and iced tea within convenient reach. All of this may have helped my concentration, but it did nothing to enhance my far from exemplary handwriting. I wish especially to thank June Hendricks, therefore, for her patience and perception in the task of typing the manuscript.

We appreciate the interested support and patience of Ruth Davison, our very good friend and housemate, and of our children, Lisa, Martha, Ruth, and Bill. Many have been the times when our children were tempted to pretend they belonged to someone else, such as when we came to a screeching halt along the highway, backed up, and collected plants from along the roadside. Although they were already grown up, there still must have been times when they might have enjoyed more attention from their parents. Often they have had to put up with their father's cooking, and on occasion have even been reduced to enduring their own.

Finally, an acknowledgment of those who have helped in the development of our book would be incomplete without mentioning the enthusiasm, encouragement, and patience shown by the staff at Indiana University Press.

Maryrose and I look forward with eager anticipation to the publication of a book that represents the help and interest of so many friends and so much concentrated effort. While savoring the satisfaction of a finished product, our greatest pleasure is the joy, excitement, and challenge we experienced while doing our fieldwork and the knowledge we have gained that increases our enjoyment of nature. Upon reviewing our lives over the past few years, we feel that perhaps we might now be ready to begin to produce a book!

INTRODUCTION

Indiana is situated in the north central part of the United States, and is also part of the region called the Midwest. It lies near the eastern border of the great interior plains. The climate of Indiana is not unlike that of the other Midwest states: on the humid side and characterized by well-marked seasonal differences, it has rather hot summers and relatively cold winters in comparison to coastal areas. That is because large bodies of water, warming and cooling more slowly than land does, have a moderating influence.

Indiana is not a particularly windy state, less so than either the plains and prairie states or the coastal areas, with an average wind velocity of between 8 and 10 miles per hour. But, as I said, it is humid, which means a generous average annual rainfall—about 40 inches per year, well distributed at a rate of 3 to 4 inches per month. More than two-thirds of the world receives less rainfall than the Hoosier State's average figures. According to the U.S. Department of Agriculture's map of plant hardiness zones, most of the northern half of Indiana is considered to be Zone 5, with an average annual minimum temperature of minus 20 to minus 10 degrees Fahrenheit. The remainder—the southern half of Indiana together with a band along the southern edge of Lake Michigan—is in Zone 6, with an average annual minimum temperature of 0 to minus 10 degrees Fahrenheit. Obviously, the wildflowers that grow in Indiana must be able to tolerate these average low winter temperatures, as well as the heat and humidity of summer. And most of the time, as every Hoosier knows, Indiana temperatures, like rainfall and wind velocity, are far from average.

A factor complicating the plant hardiness zones is the timing of warm and cold spells, particularly in Central and Southern Indiana. The Great Lakes protect the northern part of the state from drastic, unseasonal swings in temperature. Although the growing season there is more than a month shorter than in southern Indiana, when winter does arrive cold temperatures hold fairly constant until the arrival of spring. And when spring finally comes, warmer temperatures are consistent. But in the central and southern parts of the state, temperatures in February may be of sufficient warmth and duration to start spring growth of herbal plants and trees. These early periods of warmth can be followed by sub-freezing or even sub-zero temperatures much later in the spring—occasionally as late as mid-May. When this happens, dormancy is broken by the early warmth, and herbacious plants and even trees may die that would otherwise have survived the harshest Indiana winter.

From a physiographic standpoint, Indiana can roughly be divided into three sections. The northern section contains some hilly areas, bogs, and most of the natural lakes. The central and largest portion is quite flat. The southern third contains the most rugged territory in the state. Of course, a wide range of soil types can be found in each section, and so can a wide variety of native plants, all with specific needs as to soil and habitat.

The physiographic features of the northern two thirds of Indiana result primarily from the effects of glaciation. The hills in northeastern Indiana are piles of sand and gravel called moraines, which were dumped when the forward movement of the Wisconsinan glacier matched the rate at which it was melting and in effect it stood still. The lakes in this area are called kettle lakes and result from the melting of large ice blocks buried in the deep layer of sand and gravel. The northwestern section of the state was in the path of meltwater flowing south from the retreating glacier and is as a result more nearly flat.

In the central section of the state, called a till plain, thick layers of silt, sand, and gravel cover the bedrock, the results of glacial debris deposited by the melting of a glacier that had ceased to move. In presettlement times this large central area of Indiana was one of the great areas of deciduous forest in the United States, and was dominated by stands of maple and beech. These towering forests acted as a barrier to hinder the northern migration of plants with southern affinities and vice versa. While in a few places isolated stands of old beech trees still remain, these great forests were essentially lost because of the till plain's obvious suitability for agriculture. In the area of Turkey Run and Shades State Parks, the till plain has been deeply eroded well into the underlying sandstone to create the spectacular canyons found in that region. These canyon walls form a specialized habitat for many interesting plants not found in other parts of the central region.

While older glacial epochs affected the eastern and western margins of the southern third of the state, their effects have been minimized by time and erosion, leaving only thin deposits of glacial materials in upland areas. A lobe of unglaciated land extends north to Martinsville in the central part of southern Indiana. In this area physiographic regions reflect the underlying rock structure. In southern Indiana the rock strata outcrop in bands running north and south. The rock layers slope gently to the west and form the eastern part of the Illinois Basin. Until buried under the glacial debris of the central till plain, these rocks provide the rugged scenic hills of southern Indiana. Soils here tend to be badly leached and infertile. Since they are also usually acid, this is an area that has discouraged intensive farming, so more of the presettlement plant communities have been retained.

As recently as two hundred years ago, 87 percent of Indiana was densely covered by deciduous forests, with most openings being those near water—along streams and rivers, around lakes and ponds—and those created by fires, either caused by lightning or set by Indians while hunting. In the southern third of the state natural openings called glades occurred where soil covering the bedrock was too thin to support the growth of trees. These prairie-like breaks in the forest supplied a habitat for grasses, flowers, and shrubs. Elsewhere, except in early spring, before the trees leafed out, few flowers would have been in evidence.

The other 13 percent of the state, mostly in the northwest, was tall grass prairie, and there, judging from the few remnants we have managed to preserve, the wildflowers must have bloomed spectacularly throughout the spring, and in summer and early fall as well.

When the settlers came, they began clearing the land and building roads. Farmland began to replace forests and prairie. Swamps, bogs, and marshes were drained. Hundreds of non-native plants now gained a foothold in the landscape—both European weed seeds brought over by accident and garden flowers deliberately imported. Most of the "wildflowers" that we see blooming in vacant lots and fields and along roads are in fact alien species, beautiful though they may be.

But Hoosiers are fortunate that their state has not been as heavily developed as many other parts of the country. The gaudy aliens, of course, are everywhere; and throughout Indiana, but particularly in the southern third, there are unspoiled regions and varied habitats where native plants can be sought out. Identifying species one has not seen before and keeping records of blooming times and locations in Indiana can be an absorbing hobby during most of the year. It certainly has been for Maryrose and me. We have always looked forward each spring to our wildflower excursions. We have favorite sites to which we return year after year, taking our children and sometimes our friends for a leisurely trek up forested valleys and through the surrounding hills. For us these trips have become almost a ritual, celebrating the rebirth of life and hope that we associate with the coming of spring. Most frequently we go to familiar places within a few miles of home, although occasionally we are more adventurous and sample a new area we have heard about. In doing the fieldwork for this book, we were allowed to indulge and develop more fully the interest in wildflowers and their natural habitats that we have always possessed.

Although we have attempted to include representatives from the major plant groups, the selection of illustrated species in this book is totally unscientific and reflects to a large degree the flowers that we happened upon as we traveled around the state. At first we were concerned that this method of constructing a book might seem embarrassingly haphazard. Then, realizing that the flowers we were seeing were the same ones anyone touring Indiana would see, we stopped worrying about number and variety of species and painted what we saw.

This approach has worked surprisingly well for us. At the beginning of our project, Maryrose marked in one of our field guides the species that we might reasonably expect to find in Indiana. We learned early it was usually a waste of time to set out in search of a specific plant, particularly an uncommon one. As a matter of fact, several of the species Maryrose painted were inadvertently found while looking for something else!

Our prime experience in frustration was provided by a charming little plant called Spotted Pipsissewa *(Chimaphila maculata)*. I had identified a few scattered specimens at Bradford Woods when I was there with my fifth grade class for their outdoor education experience. Since that site was somewhat inaccessible and would require a special trip, I hoped to find it elsewhere. Knowledgeable spotters listed Turkey Run State Park and Portland Arch Nature Preserve as known sites. Both places were along the route on one of our trips north to Fulton County, and I was confident that we would find the plant at one or the other.

While Maryrose sat and painted in the camper in the Inn parking lot at Turkey Run State Park, I headed down the trail in search of our prize. I had been told to look under the hemlocks, most of which were not along the trail. Whichever way I went, most of them seemed to be on whichever side of the large, deep canyon I was not on. During my search I found gorgeous clumps of Rattlesnake Plantain and Wild Lily of the Valley—both of which Maryrose had already painted—but no sign of Spotted Pipsissewa. In my attempts to cross the canyon I finally found a steep trail up the fifty-foot cliff that was surely made by mountain goats—or agile adolescents who scorn the marked trails. At the top of the precipice were beautiful old hemlock trees, but no Spotted Pipsissewa. I returned, exhausted, to the parking lot, made positive sounds about the work Maryrose had done in my absence, and we proceeded north to Portland Arch.

We did not know where Portland Arch was, but had been told we could find it if we followed the signs. We found the general area with no problem. After quartering the countryside for an hour, I began stopping at the few farmhouses that showed signs of habitation and asking directions. (It seems that one of the signs had been knocked down.) When we finally arrived at the nature preserve parking lot, Maryrose resumed the work she had begun at Turkey Run while I walked the trail. The brochure in the little box at the beginning of the trail listed the points of interest. One site described was an upland area of old white pine trees—the obvious habitat for Spotted Pipsissewa! Since this was near the end of the loop, I decided that I could get the material I wanted and save time if I walked the loop backward. The white pines were beautiful, a remnant native stand. It was a perfect place for Spotted Pipsissewa—but someone had neglected to inform the Spotted Pipsissewa. Seeing nothing, I walked the rest of the loop with no better success. Back at the parking lot, I looked more carefully at the brochure and found that there are two nature trails at Portland Arch. It is likely that Spotted Pipsissewa grows on the other one, but, since the day was nearly spent and I was exhausted, we headed north toward Fulton County. This time I made no positive sounds about the work Maryrose had done while I was walking the trail.

A few days later, after beginning a couple of paintings at Bob and Laura Kern's Christmas tree farm in Fulton County, we were talking about our problems in locating Spotted Pipsissewa. Bob's response was: "Oh, that grows over in the old pine woods where we showed you the Rattlesnake Plantain last year." Since Bob was busy and I hated to reveal just how bad Maryrose's and my sense of direction was, I assured him that we knew where that was. After a careful reconstruction of our movements at the Kern farm the preceding year, we managed to get to the old pine woods. There, however, our powers of reconstruction failed. An old Christmas tree plantation, the pine woods was divided by more or less clear lanes much like the rectangular blocks in an orderly housing development. Maryrose walked a while, then went back to the camper while I continued to search the woods in what I hoped was a methodical manner. Finally I found the area where Rattlesnake Plantain grew in abundance—a much larger area than I remembered. After walking the location in straight lines six feet apart, I found what I was looking for: three beautiful little Pipsissewa plants sporting slender bloom stems topped with—nothing! Something had nipped off the pair of blossoms that should have been at the top of the stems and eaten them. After plucking a single leaf for evidence from the most robust of the three plants, so as not to stress it unduly, I returned to the camper and proved to Maryrose that I had, indeed, found the Pipsissewa. After a short conference it was decided that we were not fated to paint that particular species.

Upon completion of our project, we reviewed our original list of prospective species. Despite our misadventures, we had managed to include almost all of them. We had also found a few rare plants not listed in the general field guide. Since our purpose has been to show flowers that could be seen by people enjoying the Indiana countryside, we have tried to resist the natural temptation to seek out the rare and unusual. If a plant grows in only one or two places in the state, few people are going to happen upon it. We have, however, deliberately included a few rarities where we were fortunate enough to locate them, because we wanted to present some of the challenges that wildflower lovers savor.

We have a couple of rare plants that were included in total innocence. When in the field, Maryrose often collects and draws an interesting specimen before we do the identification and research on it. Usually this poses no problem because a plant common enough for us to find is common and distinctive enough to easily identify. One such instance involved a large, beautiful Lady's Tresses orchid. When I tried to identify it, I found that there were several species so similar that I was not comfortable assigning it a name. Upon referring it to knowledgeable authorities, we found we had dug the only officially verified specimen of *Spiranthes vernalis,* previously unknown to be found in the state.

Another type of "mistake" occurred in Plate 40 which was reserved for Featherbells. We had located a large, beautiful colony of Featherbells in Fulton County, but had nothing to put with them. Since Maryrose likes to include several species in each painting, we were discussing hypothetical combinations of plants that could grow in the same habitat as Featherbells. We were traveling Highway 31 north of Indianapolis at the time, and I remembered that I had seen an interesting plant near Westfield the preceding year but had not stopped to examine it. We found the site, collected a specimen, and continued on our way to Fulton County and the Featherbells. We conjectured that the plant we had found—which turned out to be *Althea officinalis* or Marsh-mallow—could conceivably grow in the same habitat as the Featherbells. It was only after the preliminary work on the painting was completed that we found there are only two known sites for Featherbells in the state, and Marsh-mallow grows at neither. In fact, it appears that Althea officinalis was not previously known to grow in the state at all. So, while it might hypothetically be possible for Featherbells and Marsh-mallow to grow together in Indiana, we know it isn't so!

While we have taken great care in the accuracy of our identifications and descriptions, our purpose has not been to write a field guide. Instead, we hope to share an increased awareness and appreciation for the peace and beauty of nature that we have experienced during the course of our work.

Awareness of the plants that grow around us in such abundance and variety is an aptitude that can be enhanced by practice and experience. When Maryrose first started to specialize in the painting of wildflowers, our experience had been almost entirely with those woodland plants that bloom in early spring. We were familiar with only a few common summer flowers. The finding of Wild Senna (Plate 63) along Brummett's Creek Road a few summers ago demonstrated to us just how poorly developed our ability to perceive really was. A few days earlier, I had been looking through a book showing photographs of wild plants that adapt well to cultivation and happened to notice Wild Senna. It was totally unfamiliar to me and I was impressed by its unusual appearance. Our surprise and pleasure at finding it growing in rank clumps beside Brummett's Creek Road were tempered by the guilty realization that we must have passed its distinctive blossoms many times over the years without seeing them. After sighting that first plant, we have since seen it in many other areas.

The problem of being unable to see a plant until we have identified it and learned to recognize its shape in the landscape has in several instances led us to false assumptions about a plant's range. It was in northern Indiana that we first identified Button Bush (Plate 47) and Germander (Plate 52). Since we did not recall ever having seen these plants before, we assumed that they were northern plants. Subsequently, to our chagrin, we have found that both are fairly common all over the state. They both grow in areas we were familiar with. We had seen them before but had not perceived them.

In this book we have arranged the plates in order of season to provide a progression of bloom from earliest spring, when snow might still be found in isolated patches, to fall days, when all but the most adaptable plants have been killed by the frosts that announce winter's approach. The dates given are those of the actual beginning of the paintings. So, while we can affirm that the species shown in a painting were in bloom on the date given, that date may be truly valid only for the year and location illustrated in the painting.

Having completed the paintings for the book over a period of several years, we found that conditions of temperature and moisture have a profound effect on the bloom times of wildflowers. These variations can cause bloom time to be more than a week earlier or later than the norm. When a season of unusual lateness occurs, plants that would normally have dropped their petals and be developing seedpods may be seen blooming in unusual combinations with later blooming flowers. Often such a fluctuation is not consistent throughout the course of a year. A late season of spring bloom may be followed by unusually early summer bloom if growing conditions are sufficiently dry. The problem of assigning accurate bloom times is further complicated by the difference in the length of the growing season in southern and northern sections of the state. Spring comes later and fall earlier in the north, so that flowers that commonly follow each other in the southern areas bloom together in the north.

Flowers that bloom in spring tend to be fairly simple in structure and to have a short period of peak bloom, sometimes lasting only a few days. Spring blooming plants are usually perennials, dependent upon food stored in tubers, corms, or bulbs to supply the quick burst of energy needed to expand their leaves and flowers. These plants have usually bloomed and produced their seed by the time the leaves on the trees are fully expanded. Some die down quickly, while others persist until midsummer with leaves dramatically enlarged to utilize the greatly reduced light reaching the forest floor through the dense canopy of tree tops. Flowers are sparse in the deep forest once spring has passed.

Most flowers of summer and fall bloom in bright areas along the edges of forests or in the open, exposed to full sunlight. Bright sunlight, warm temperatures, and abundant moisture produce dense, lush growth in Indiana during the summer months. Plants tend to grow tall in their competition for light. Flowers have had time to develop complexities absent in most spring blooming plants. Probably the most common of the summer and fall flowers are the daisy-like blossoms. Although spring flowers such as Bloodroot and Hepatica have a superficially daisy-like shape, they are only single, simple flowers without a central disc. Summer blooming flowers of this general appearance are a composite of numerous tiny flowers—sometimes more than a hundred—that form the central disc and petals. Each petal is a separate flower, and the button-like central disc is made up of concentric rings of tiny flowers that begin blooming at the outer ring and progress to the center. This complicated construction is one of the reasons that daisy-like flowers tend to be long lasting.

While many summer and fall blooming plants grow from perennial crowns, many annuals have had time to germinate and grow to flowering size by midsummer. Both annuals and perennials that flower during this time of year tend to have long seasons of bloom. Some begin in late June or early July and extend into September. As with most generalizations, of course, there are rather startling exceptions. An example is Nodding Pogonia or Three-birds Orchid

(Plate 66), which blooms in August for an even shorter time than the spring flowers, often only a day or two.

Sometimes, in a favorable growing season, flowers such as Black-eyed Susans and Butterfly Weed may have a secondary bloom. The heavy primary bloom begins early when the mature plants flower. A lighter, secondary flowering may come later when first year seedlings attain enough size and strength to bloom in their first season. Occasionally delayed bloom occurs when a plant is damaged either naturally or artificially by mowing. When cut short the plant often sends out secondary branches that bloom when they are sufficiently developed. The surprising variation in the times plants bloom has supplied a never-ending source of conversation as Maryrose and I have toured the state over the past several years.

Timing and duration of bloom season were factors of great concern to Maryrose as she planned her painting schedule. Each painting required from ten to fourteen days of intensive work for completion. The length of bloom season for many concurrently blooming plants was so short that by the time she finished painting one the others would be out of bloom. This problem was solved by planning the paintings carefully, drawing the subjects, and doing finished painting on small areas of the drawing to establish color and give the information about texture and plant structure that would be required to finish the paintings at a later time. For most of the paintings this process took about two days. At the end of each painting season Maryrose's portfolio was full of embryonic paintings that looked as if they had the measles. The illustration boards were spotted with bits and pieces of flowers, stems, and leaves surrounded by notes scrawled in the margins. During the winter months she would work at home in her studio, filling in the rest of the paintings and adding the backgrounds.

Over the years Maryrose has developed her own watercolor technique specifically for the painting of plants and flowers. Realism is achieved by an excruciatingly slow process of applying several successive layers of transparent watercolor. At the time of application the paint must be as dry as possible, since too much water would cause the layers beneath to dissolve and blend together. This process produces much greater vibrancy of color and a greater range of subtle surface variations than mixing a color and applying it in a single layer, and it produces a painting that retains its integrity even when greatly magnified. Maryrose herself hadn't realized how much this was so until she used a detail slide at one of her presentations. The slide, which showed only a couple of inches of the painting, was projected to cover an entire wall many times larger than the screen she normally uses. The stick and fern frond shown in the slide still looked detailed and realistic even at that magnification.

Maryrose is not happy unless she has a fresh, living specimen to work from. She needs to examine texture and view her subject in three dimensions. For most of the paintings we lifted specimens by cutting a plug of soil with a sharpened spade and fitting it into a plastic nursery pot. We found that plants handled in this manner could be placed back in their original location after Maryrose was through working with them. If given a thorough soaking to reestablish contact between the plug and the surrounding soil, they showed no signs of noticing they had been disturbed. When plants were too large to dig we cut a stem early in the morning or late in the evening and put it in water. In some cases Maryrose took her paints and drawing board and worked at the growing site of plants that were too rare to risk disturbing.

The only painting in the book that was done from slides is Pink Lady's-slipper. Maryrose had intended to paint this rare flower on location. The only place we were able to find it in Indiana was at a nature preserve where a boardwalk has been built to provide access to a sphagnum bog. When we arrived, we decided to first examine the site to select the species we wanted to include with the Pink Lady's-slipper and to locate the specimens for the painting. At first, we strolled back and forth along the boardwalk in awestruck wonder. Here were splendid specimens of plants we had read about and seen in pictures but had never before encountered actually growing. This was our first experience with the beautiful Tamarack tree with its artistically shaped branches bedecked with tufts of blue needle-like leaves. Our surroundings were so alien to our previous experiences that, except for the boardwalk, we might well have found a time machine and traveled back to an earlier epoch. The visual magic of the place was further enhanced by a symphony of sounds from the abundant birds, insects, and frogs that enjoyed the same protection given the rare plants. But all too soon we began to be aware of the discordant element that prevented Maryrose from doing any preliminary work at the site. Mosquitoes, singing off key, had summoned their friends and relatives and were giving a concert that we didn't want to hear—and repellant is something we always seem to have on hand except when we really need it. By the time I got around to photographing the specimens to use as subjects I was twitching so badly that I had trouble holding the camera steady.

One of the concerns we had at the beginning of our project was producing paintings that would be representative of the different areas of the state. We considered doing a painting from each of the counties, but found that the wildflower population is unaware of political boundaries. It was meaningless to attempt to select plants that were representative of areas as small as counties. We decided that use of the Indiana park system was our best vehicle for presenting a representative sampling of the different regions and habitats of the state. Indiana parks have been planned to provide reasonably convenient access to residents in all parts of the state, and as a result the parks are fairly evenly distributed.

We were given written permission from the Department of Natural Resources to dig specimens of common plants in the state parks on the condition that we notify the property managers when we were working in their park. We reported at the end of each season the species names of the plants we had dug. We were expected to be discreet in our methods of obtaining specimens and to replant them when we were through. Actually, we felt like criminals much of the time.

Our usual method of operation after checking in at a new park was to drive all the park roads very slowly, examining the roadsides to spot species that we might want to include. When that source had been catalogued we would spend a portion of each day walking the trails. After deciding upon sets of plants that could reasonably be combined, we started collecting our specimens. Most of our collecting was accomplished early in the morning. We found that plants lifted or cut then suffered less stress, and we were less likely to encounter other people. When collecting along the park roads I dug while Maryrose watched for the approach of hikers, cyclists, or motorists. If anyone appeared on the scene I held my short handled shovel and plastic pot behind me or dropped them in tall grass and tried to look as though I were innocently smelling the flowers. Most passersby probably thought I was answering a call of nature. We usually collected only one or two plants at a time so that they would retain their fresh, natural look while Maryrose drew and painted them. Also, Maryrose found it frustrating to have too many plants waiting to be worked with at one time.

After selecting the species to be combined in a painting, the most enjoyable part for Maryrose was putting them together in a natural looking, pleasing composition. She is often asked if she found her specimens growing as she has shown them in her paintings. The answer is yes and no. Although she often finds placements and shapes in nature that suggest the composition of a painting to her, she seldom if ever has found a plant grouping that can be drawn and painted without modification. In nature, plants tend to have more space around them than can be allowed in a painting—particularly plants of different species. This is especially true of spring blooming plants. Another problem that concerned her was presented by summer plants, many of which bloom in narrow, vertical spikes. Variation in composition is difficult when combining flowers that all have essentially the same shape. She addressed this problem by adding a background scene that tied the painting together and gave information about places where the plants shown might be found.

Using the Indiana park system as our base of operations has opened a whole new set of experiences for us in the enjoyment of nature. Our previous use of the park system had been limited to family gatherings at the picnic areas or casual sightseeing along the park roads. While we were vaguely aware that parks had hiking trails, we were totally ignorant of their value in observing wildflowers specifically and nature in general. These trails, enjoyable to those who like to walk simply for exercise, provide marvelous opportunities for observation to people who are interested in the plant and animal members of natural communities. The trails are well planned to provide access to a variety of interesting areas, not only woods but also meadows, wet lowlands, and other environments. Maryrose discovered the richness and scope of this resource first and gave me glowing, enthusiastic accounts of her treks. My initial reaction was skepticism, thinking that public trails must be barren, worn-out places compared to the wilderness areas broken only by game trails that we were accustomed to enjoying on private land.

My first experience with a park trail changed my mind quickly. The trail itself was broad and smooth, with a top dressing of wood and bark chips in areas that were often wet. Beyond its margin lay undisturbed forest floor thickly carpeted with, in this instance, the foliage of Squirrel Corn and Dutchman's Breeches. I had never seen such lush growth even in my favorite "out-of-the-way" places. Careful inspection of the trail sides showed that variety was present as well as volume. There were several species that I considered choice and uncommon.

When Maryrose first started doing fieldwork in the state parks we dug out the old family tent. This miracle of lightweight efficiency was the relic of a six-week camping trip we—three adults and four children in a station wagon—had taken some years ago. Aside from a bit of mildew, the tent had survived relatively intact. After surprising ourselves by being able to find all the poles and parts we decided that we needed to be sure that Maryrose would be able to put the tent up by herself before setting out for unknown territory.

Choosing a warm sunny weekend early in the spring for the experiment, Maryrose and I, along with several of our children and friends, headed to a nearby park. During the course of the afternoon we wondered what the passersby were thinking as they observed the frantic efforts of one red-faced woman to get one part of a two-room, 11 x 16-foot tent up before the other parts fell down. We all felt we were probably making a bad impression sitting around in our lawn chairs, chatting, and eating snacks while all this was going on, and considered getting out of sight until the job was accomplished but were afraid Maryrose might miss our constructive comments of advice and encouragement. She found that the tent could be erected by one person, but the challenge was such that we tried to time tent moves so that help would be available.

At first Maryrose camped in parks close to home. She stayed in primitive camping areas whose sole amenities were cold water spigots and pit toilets. I stayed with her on weekends and checked with her each day by phone at a predetermined time. In the beginning she felt extremely insecure sleeping alone in a tent in a deserted campground, but soon became so confident that she stopped keeping a hatchet under her pillow. Actually, we felt fairly comfortable with the camping because the parks are so well patrolled at night. In addition to the family tent, we had resurrected an old gasoline campstove and lantern that had belonged to Maryrose's parents and bought an ice chest. With these, we were able to eat well and live in reasonable comfort. We noticed that after a few days in the field a dishpan of warm water made us feel as clean and refreshed as a shower at home. The old saying that everything is relative is apparently true.

We adapted rather well to this primitive living, coping with recalcitrant air mattresses, pesky insects and animals—especially raccoons—and the elements, mainly heavy rains and strong winds. In the process, we collected a rich store of memories, vivid out of all proportion to the length of time involved, for this phase of our fieldwork lasted only a couple of months. Its demise was sealed when our teenaged daughter accompanied Maryrose and me on one of our northern trips. She asserted that modern facilities and electricity were indispensable ingredients to her grooming and comfort. This meant a Type A campsite instead of the Type C ones we had been frequenting. As we reassessed our camping techniques, we decided that so long as we were going to pay for electricity, we might as well enjoy it. It occurred to us, for example, that the tiny refrigerator the kids had used in college could be pressed into service and probably compensate for the difference in campsite cost by eliminating the need to buy ice for the cooler, and an electric fan would certainly be welcome during hot weather. It was not long until we added a television, coffee maker, hot plate, electric skillet, and spotlight, and were on our way to being thoroughly corrupted.

We felt close to nature while tenting and treasure the memories of our experiences. But an inordinate amount of time and energy was expended simply coping with the elements and the animals. Moving the tent from one park to another was an all-day job that always seemed to be necessary when the temperatures were in the upper nineties. Losing one day moving our equipment and most of the next recovering from near heat stroke finally made us decide to invest in a used pickup truck and slide-in truck camper. Of ancient vintage, our camper was nevertheless clean, comfortable, and sturdy. Best of all, it was water and raccoon proof! On our first test run, we spent most of the first hour wandering aimlessly about in wonder-stricken awe—setting up camp had consisted of plugging in the extension cord connecting the camper to the electrical outlet at the campsite.

While our fieldwork experiences have culminated in the convenience, comfort, and efficiency of modern recreational living, we wouldn't for the world have missed our two years of tenting. Even stressful situations are remembered in retrospect with satisfaction, such as the night Maryrose spent alone in a near tornado, serving as a sturdy extra tent pole in her efforts to keep the tent from blowing away. Her experiences have made her feel that she was given a challenge and persevered. Most of all, we have enjoyed the opportunity to do together work that we both love. So unhesitatingly, then, we can say that writing and illustrating *Wildflowers of Indiana* has been rewarding from start to finish.

WILDFLOWERS OF INDIANA

Plate 1

The first plate shows only one species—Skunk Cabbage *(Symplocarpus foetidus)*—for the simple reason that Skunk Cabbage blooms much earlier than the other plants growing in its habitat: the shell-like spathes begin emerging through the forest litter in February. Cold spells and late snows do not intimidate this interesting plant; it produces enough heat through biochemical processes to melt its way through ice and snow, and its spathe provides a haven for its pollinators that is many degrees warmer than the surrounding temperatures.

The two to five inch purple spathe is mottled or streaked with lighter colors. Pointed at the tip, it is open on one side to expose the short, knob-shaped spadix which it protects. Distributed over the spadix are tiny yellowish flowers that, when pollinated, produce seeds that grow under the spongy surface of the spadix. At bloom time, leaves begin to appear beside the spathe. At first they are tightly coiled into small cones. A single plant produces six to eight egg-shaped leaves that may reach three feet in height and a foot in width by the end of its growing season. These plants usually grow in colonies and may completely fill an area of suitable habitat with their huge stems and leaves. By the end of the growing season the spadix has decomposed, leaving a pile of pebble-like seeds on the surface of the ground.

Skunk Cabbage has large, deep roots and is long-lived. Its common name comes from the cabbage-like appearance of the leaves and the fetid odor that is noticeable when any part of the plant is crushed. All parts of the plant contain calcium oxalate crystals, which burn the mouth and throat if eaten fresh. When thoroughly dried, however, the leaves may be used in soups and stews, and the roots may be ground into a pleasant-tasting flour something like cocoa.

The specimens for this painting were found growing in a deciduous woods at Pokagon State Park, in black peat saturated with water. The area was covered with dry leaves and seemed firm, but the ground trembled as we walked. It was a bog—a place that in my mind is associated with strongly acid soil. The calcareous nature of this bog's soil was revealed by the numerous snail shells we found in the area. The acid-producing conditions of continuous saturation and decomposition of organic forest litter were neutralized by the lime present in the soil.

Maryrose usually works at her drawing table with specimens that have been dug or cut. Since Skunk Cabbage did not look like anything amenable to either digging or cutting, Maryrose sat on a large plastic bag and did the drawing and initial painting at the site. While she worked I wandered around, exploring and taking pictures of Skunk Cabbage to be used later for reference. But I kept checking on her as she drew, expecting her to sink slowly out of sight into the black depths of the bog. My apprehension was totally unfounded, I am glad to say.

March 6, 1987

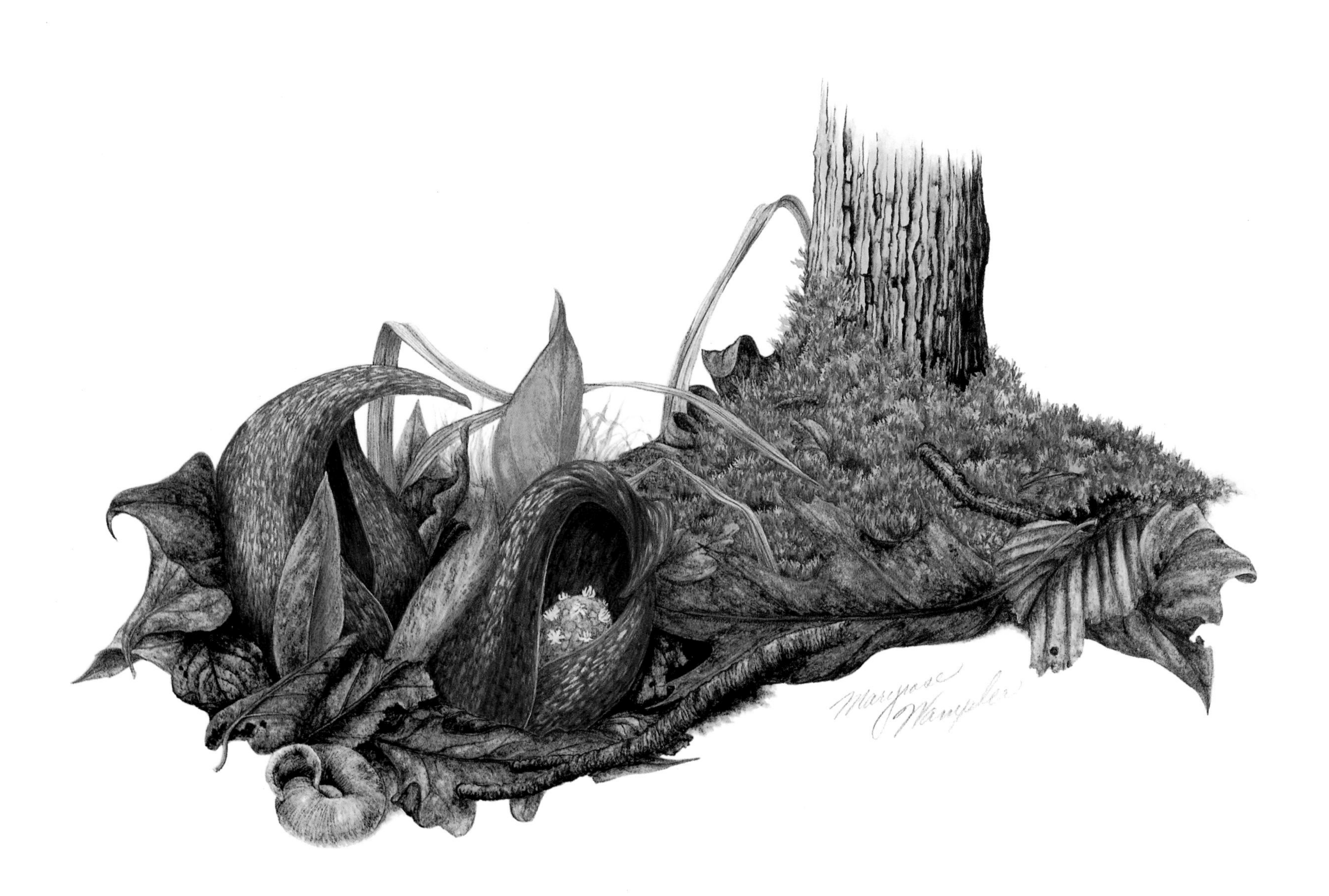
Maryrose Wampler

Plate 2

Plate 2 shows the fragrant, pinkish flowers of Trailing Arbutus or Mayflower *(Epigaea repens).* Clusters of one-half inch, five-lobed, tubular corollas are borne in leaf axils and at the tips of branches. The evergreen, alternate leaves are three-fourths to three inches in length and are oval in shape. Arbutus is actually a trailing, woody-stemmed shrub growing only two to three inches high with branches eight to fifteen inches long. Both stems and the undersurfaces of leaves are covered with stiff, rust-brown hairs. In favorable habitats it forms dense mats of foliage. The plant spreads by means of shallow underground stems and has a sparse root system that renders it extremely difficult to establish artificially. For this reason, naturally growing plants should never be disturbed. Arbutus can be propagated without threatening the parent plant by taking short stem cuttings in the fall and rooting them in much the same manner as cuttings from rose bushes.

After flowering, Trailing Arbutus produces small white berries that when ripe split into five sections, exposing whitish pulp covered with tiny seeds. The berries are enjoyed as food by both ants and birds, which aid in their distribution.

Trailing Arbutus grows in open pine or oak forests where there is good drainage and acid soil. It favors exposed ridges where the litter of fallen leaves does not accumulate to a depth great enough to smother its growth. This beloved wildflower is most common in the Appalachian mountains and associated uplands in the eastern part of the United States. Because of its spicy fragrance, beauty, and lasting qualities as a cut flower, it has disappeared from many areas due to over-picking.

Trailing Arbutus is found in a few scattered locations in Indiana. The specimens shown in this painting were found in an area east of Bloomington that has attracted the attention of wildflower lovers since pioneer days. This particular site is of special interest to me because my father lived as a boy, some eighty years ago, at the foot of the hill where the arbutus grows. He recalled that even in those days students and teachers from Indiana University made a pilgrimage each spring to see it. Its preservation is perhaps due to its relative inaccessibility and to the interest that knowledgeable local people have taken in it.

These small secluded patches of flowers are held in great esteem in Indiana. Trailing Arbutus is the official flower of Indiana University; the yearbook is named *The Arbutus;* and the Jewel of Office worn by incoming presidents of the university at their inauguration ceremonies prominently features this charming wildflower.

Partridgeberry *(Mitchella repens)* is shown growing with the Arbutus. Favoring a similar habitat, this creeping evergreen grows even closer to the surface of the soil. The shiny round pairs of dark green leaves and the bright red berries that persist all winter (if not eaten by birds) make this a popular Christmas decoration. A splendid groundcover, Partridgeberry adapts readily to moist, shaded areas. In June and July small fragrant white flowers bloom in pairs in the leaf axils. The two flowers join to form a single berry.

Lichens are common companions for Trailing Arbutus and Partridgeberry. Symbiotic combinations of algae and fungi, these interesting plants are able to withstand dry periods by shriveling and going dormant, then quickly returning to active growth when moisture is available. Common names for the two types shown are Pixie Cups and British Soldier Lichen. The Pixie Cups are shaped like shallow goblets, and are often seen holding a droplet of water. The British Soldiers have red caps atop columnar shapes that suggest a group of soldiers clustered together.

March 8, 1987

Maryrose Wampler

Plate 3

Specimens for Plate 3 were found in Monroe County south of Bloomington. The white one and a half to two inch flowers of Snow or Dwarf White Trillium *(Trillium nivale)* look comparatively large on the tiny two to six inch plants. The flowers are held above the three oval, blunt-tipped, one to two inch dark green leaves. The smallest and earliest of the genus, Snow Trillium is sometimes found blooming while patches of snow are still on the ground.

Snow Trillium is one of those plants that can be quite numerous at small sites, but is not always easily found. The year before Maryrose painted Plate 3, we were visiting Cedar Bluffs, a protected nature preserve. An area of limestone cliffs overlooking a stream, Cedar Bluffs supports a great variety of wildflowers. Several different habitats are well represented in a small space. A rich, forested alluvial valley lies along the stream at the base of the bluff. The more or less vertical limestone walls have breaks and pockets of soil; they are a hanging garden for plants that need sharp drainage but enjoy a constant supply of moisture seeping from the rock strata. At the top of the bluff is a prairie-like habitat where a thin layer of well-drained soil covers areas between outcroppings of bedrock. In this upper area we found the foliage of snow trillium—our first sighting of this diminutive beauty. The blossoms had long since withered and seed capsules were already forming.

Since we didn't have permission to collect specimens at Cedar Bluffs in any case, we explored several sites on our way back home. The road followed the ridge overlooking the stream for a distance and we found a large colony of plants where the road ran near the brink of the bluff. At a nearby farmhouse we were assured we would be welcome to collect a few plants when they were blooming.

The next spring we returned at the proper season and searched the area carefully—we thought! After passing the crest of the hilltop we were examining, we looked back up the slope and saw that we had passed a colony of hundreds of flowers, almost stepping on some of them. The blossoms were all facing the same direction, so they had escaped our notice until we got a full-face view. Later we heard that this little gem is found in large numbers growing along the limestone bluffs that border Cascades Park in Bloomingon.

The tiny flower in the center of the painting is Harbinger of Spring or Salt-and-pepper *(Erigenia bulbosa)*. Growing from a round tuber deep in the soil, this plant has one or two leaves divided into narrow oval or lobed segments. Small leafy bracts are found at the base of the umbel of flowers. The white flowers have prominent, red-brown anthers. Harbinger of Spring is a name merited by the early bloom season; it is always one of the very first wildflowers we see each year. The name Salt-and-pepper reflects the appearance given by the dark of the anthers contrasting with the white of the petals.

April 1, 1986

Plate 4

Little-used county roads often follow the relatively flat valleys of small streams. Steep, natural wooded slopes bordering such roads provide settings of great beauty for a wide variety of spring wildflowers and ferns. Their habitat is well drained, rich in humus, and adequately supplied with water that gradually seeps from the layers of bedrock exposed along the faces of the hills. The specimens for this painting came from flowery hillsides bordering a wooded country road in Monroe County. Here, trickles of water glide over bare rock strata exposed in the rills of a steep hillside, and festoons and cushions of moss peep out among drifts of dry leaves. Colonies of flowers spring from pockets of soil caught between the rocks. Miniature waterfalls formed by resistant ledges of rock splash into tiny basins, providing aural as well as visual pleasure.

This is a typical setting for Hepatica *(Hepatica acutiloba)*. The three-lobed basal leaves persist through the winter and become mottled and weathered in appearance though still living. These year-old leaves vary in color from green to maroon. When the flowers bloom in early spring, last year's leaves cascade down the slope on their thin four to eight inch stems, providing a beautiful foil for the fuzzy-stemmed one inch flowers that bloom in white, pink, or blue profusion from the crown of the plant. As the season progresses, the plant becomes even lovelier when the new leaves make their silvery, silken appearance at the base of the flower stems. There are several days toward the end of the bloom season when old leaves, new leaves, and a nosegay of flowers are all present at the same time.

In the past Hepatica has had medicinal uses. People used to believe that when the shape of plant parts suggested the shape of body organs or members, the plant would cure ailments of that particular body part. This belief is referred to as the Doctrine of Signatures and forms the basis for much of the herbal medicine in medieval times. The three lobes of the Hepatica leaf, it was thought, resembled the three lobes of the human liver. The generic name comes from the Latin word for liver, and the plant is sometimes called Liver-leaf.

There is another similar species *(Hepatica americana)* that differs only in having a more rounded lobe on the leaf. There are also intermediate forms where the two species have hybridized. Acutiloba, in my experience, is the more common species in Indiana.

Rue Anemone *(Anemonella thalictroides)* is a delicate plant whose flower resembles that of Hepatica although the color is usually white or white lightly blushed with pink. Both flowers lack true petals. The petal-like structures are actually sepals with green bracts behind them. Several stalked flowers rise above a pair or whorl of dainty, three-lobed leaflets. The leaves are an unusual reddish green color that provides a setting of fairy-like delicacy for the flowers.

The leaves of the plant resemble those of the Meadow Rues. This similarity supplies the Rue portion of the common name and the thalictroides in the scientific name.

A single specimen of Spring Beauty (discussed in detail in Plate 6) is at the center of the painting. This plant shows the pair of leaves halfway up the stem and the initial flower blooming at the base of the curled cluster of buds.

April 5, 1984

Marjorie Wampler

Plate 5

Plate 5 depicts three species of early spring wildflowers: Bloodroot *(Sanguinaria canadensis),* Cut-leaved Toothwort *(Dentaria laciniata),* and Twinleaf *(Jeffersonia diphylla).* These wildflowers commonly grow together on wooded slopes, such as are found at Clifty Falls State Park. The small plant in the foreground is Fragile Fern.

One of our showiest wildflowers, Bloodroot can be found in large colonies. The sparkling white one and a half inch flowers are borne one to a stem. When it first emerges from the ground, the flower bud is enclosed in a single silvery green leaf that curls around it protectively. As the plants develop, the flower stems elongate first and raise the flowers above the level of the leaves. At this stage the flowers are prominent, and a large colony is a pool of pristine white jeweled with prominent golden stamens at the center of each flower. The flowers open wide when the sun strikes them through the still bare trees and close when evening comes.

As their growth cycle progresses, the leaf stems also elongate and the leaves expand and become a dark green. With the growth of the leaves the flowers appear more widely spaced and gradually disappear as their petals drop, exposing the oval, pointed seed capsule. By early summer the Bloodroot colony is a lush patch of leaves six to seven inches across and up to ten inches tall. When the seed capsules mature, they split open, allowing the seeds to roll down the slope to extend the perimeter of the colony.

Both the common and scientific names refer to the abundant red-orange juice contained in the root and stem of the plant. This bitter-tasting sap was used by the Indians as war paint on their bodies and as a dye for clothing and baskets. Mixing the Bloodroot sap with tannin from oak bark sets the color and makes it more permanent.

Bloodroot is a member of the Poppy family, and both the plant and the tuber are poisonous. The plant contains alkaloids that are closely related to morphine. Medicinal uses made of it by Indians and early settlers were related to relief of pain. Because of its poisonous nature, most applications were external.

Cut-leaved Toothwort has a terminal cluster of six to twelve white, four-petaled flowers that blush pink as they age. The two to five inch leaves are deeply lobed and sharply toothed. They form a whorl of three leaflets above the middle of the stem. The common name derives from the tooth-shaped rhizomes that lie deep in the forest soil. Another common name is Pepper Root, derived from the mild, horseradish-like flavor of the rhizome. Early settlers collected the roots of this plant to flavor soups and stews.

Twinleaf is a short-lived flower about one inch across that resembles Bloodroot. Its interesting seedpod is shaped like an upside-down pear with a lid at the top that opens with a hinge-like attachment when mature. When the plant first blooms, the flower stem is longer than the leaf stem. As the season progresses, however, the leaf enlarges greatly and grows twelve to eighteen inches in height. The two wing-like segments that comprise the single leaf give the plant its name. The generic name, Jeffersonia, refers to Thomas Jefferson and was given by a botanist friend of the third president. The leaves resemble butterflies in flight and are beautiful throughout the life cycle of the plant. This flower forms large colonies and may cover an entire woodland hillside.

April 7, 1984

Plate 6

The wooded hillsides along Brummett's Creek valley in Monroe County have for many years been the site of one of our favorite spring wildflower excursions. Here a fairly wide variety of spring flowers—including those seen in Plate 6—grows in great profusion in the rich humus of the mixed hardwood forests that densely cover the steep hills.

In the left foreground is Spring Beauty *(Claytonia virginica).* Its one-half to three-fourths inch flowers are pink or whitish with dark pink stripes. The flowers are borne in a loose raceme on a rather spindly stem six to twelve inches tall. They close up when picked or when the sun is not shining. There is usually a single pair of succulent dark green linear leaves midway up the stem. Since the lower part of the plant is frequently masked by a loose layer of dry leaves and forest litter, the colonies of flowers give the appearance of rising above a fountain of basal leaves. These flowers can grow in such profusion that a hillside may appear to be dusted with snow. Sometimes they are found thriving in shaded lawns. Their deep tubers and short, early season of growth allow them to carpet a lawn without damaging the grass. The small potato-shaped tubers have a sweet flavor like chestnuts and were eaten by both Indians and early settlers. They are still prized by people who collect wild foods.

The Downy or Woolly Blue Violet *(Viola sororia)* is shown above and to the right of the Spring Beauty. Its light blue flowers bloom on four to six inch stems that are densely covered with short, soft hairs. The broadly heart-shaped leaves are also hairy on their undersides. They are found in open, well-drained woods.

The cup fungus growing on the root of the dead stump is probably *Urnula craterium.* This interesting fungus sometimes appears to release a wisp of smoke when its spores are ejected to drift to a new location on the forest air currents.

The popularity of the yellow flower at upper left is indicated by its numerous common names. Variously called Trout Lily, Fawn Lily, Adders-tongue, and Dog-tooth Violet, *Erythronium americanum* is familiar to almost anyone who has ever taken an early spring walk through the woods. Trout Lily and Fawn Lily refer to the mottled pattern on the leaves, Adders-tongue to the shape of the immature seed pod, and Dog-tooth Violet to the appearance of the small bulbs that form on the network of underground stems. A true lily, the plant has a one to one and one-half inch flower that blooms at the top of a slender, leafless stem and is composed of three reflexed sepals, often with a dusting of cinnamon on the outside, and three narrower petals that also curve back. These plants must be several years old before they bloom. Blooming plants have two glossy, elliptical leaves six to eight inches long mottled with maroon or silver. Young plants have only one leaf. Often dense colonies are seen with a few large, blooming plants surrounded by hundreds of young, single-leaved plants.

The white form of the Trout Lily *(Erythronium albidum)* is shown to the right of the stump. It is similar in growth habit to the yellow lily, but tends to have narrower, lighter colored leaves. Sometimes there is a shading of lavender on the outside of the white flowers.

Between the two Trout Lilies is Squirrel Corn *(Dicentra canadensis),* with flowers that are heart-shaped and carried several to the stem above the cluster of ferny foliage. The name comes from the root tuber that is shaped like a kernel of corn.

The flower below the White Trout Lily is Dutchman's Breeches *(Dicentra cucullaria),* whose white or pinkish pantaloon-shaped flowers droop in a row from the leafless, gracefully arched stem. The long stalked basal compound leaves are much divided and give the plant a light, feathery appearance. At their base is a cluster of many small white tuber-like roots.

Squirrel Corn and Dutchman's Breeches are closely related to each other and also to the cultivated old-fashioned Bleeding Heart. Plants that are not blooming can hardly be told apart without examining the tubers. The generic name means "two spurs," and the two flowers are similar in structure although the shape of the spurs is different. These flowers are pollinated by bumblebees. Honeybees can gather the pollen but cannot reach far enough into the flower to collect the nectar or fertilize the pistils.

The sponge fungus is *Morchella esculenta,* the highly prized edible morel mushroom. We have found these delicious morsels in large quantity only once. Usually we find only enough for Maryrose to use as painting specimens. Unfortunately, that process is so time-consuming that by the time she is finished with them, they are too dried up to eat!

April 10, 1984

Plate 7

The flowers shown in Plate 7 can be found along the hills overlooking the Whitewater River near the town of Metamora. The two inch bright yellow flowers of Wood Poppy or Celandine Poppy *(Stylophorum diphyllum)* grow in rank profusion in this area. The large, showy blossoms are composed of four broad, round petals cupping a single pistil and numerous golden stamens. The nodding ovoid buds are covered with two fuzzy green sepals that drop off when the flowers open. Flowers are solitary or in small clusters at the top of a stem that has only two paired leaves, thus the species name diphyllum, meaning two leaves. Both stem leaves and basal leaves are pinnately divided into deeply lobed, rounded segments. Leaves are pale green with a whitish bloom on their undersurface. The hairy, egg-shaped seedpods are similar to the buds in appearance.

Celandine Poppy makes an excellent addition to the wildflower garden where conditions of moist organic soil and deep shade prevail. The plants are easily moved or can be started from collected seed. Celandine Poppies remain densely foliaged in low, neat-looking mounds well into the summer. Unusual for a spring blooming wildflower, blossoms occasionally appear long after the main bloom season is past. If protected from encroaching neighbors, Celandine Poppies will persist for many years and form a lovely groundcover for deeply shaded areas.

The plant resembles and is sometimes confused with Celandine *(Chelidonium majus),* but Celandine has smaller flowers, alternate leaves, and a smooth, slender seedpod. It is a somewhat weedy plant from Europe that has naturalized.

The unusual flowers of Blue-eyed Mary *(Collinsia verna)* are shown across the top of the painting. Whorls of four to six stalked blue and white flowers arise from the axils of the opposite, mostly sessile leaves. The one-half inch flowers are composed of two lips, with the upper lip divided into two white lobes and the lower into three blue ones. The middle lobe of the lower lip is folded backward to envelop the stamens and style and so is hidden. This weak-stemmed annual grows six to eighteen inches tall and has one-half to two inch ovate leaves. The seeds germinate in open, damp woods in the fall and over-winter as young plants. The blanket of fallen leaves held in place by grass and herbaceous stems killed by fall frosts is necessary to the well-being of Blue-eyed Mary, for it protects against not only drying winds and extreme cold, but the rapid changes in temperature that produce frost heaving.

Missouri Violet *(Viola missouriensis)* is a small plant that bears pale, white-centered, lavender flowers of typical violet shape. Flowers and leaves are borne on separate stems that arise from the crown of the plant. The plant holds its blossoms above the triangular leaves.

April 10, 1986

Maryrose Wampler

Plate 8

Gill-over-the-ground or Ground Ivy *(Glechoma hederacea)* is shown in the foreground of Plate 8. Of European origin, this perennial has naturalized widely. Used primarily for groundcover in damp, shady places, it can grow in full sun if sufficient moisture is present. Because of its creeping growth habit it withstands close mowing and can become invasive in lawns. When Gill-over-the-ground is mowed, a strong catnip-like fragrance is noticeable. It is an attractive little plant with opposite scalloped, round leaves. The purplish-blue flowers are about three-fourths inch long and have a narrow upper lip and a broad lower lip spotted with purple, with three spreading lobes.

The bright yellow flowers of the Dandelion *(Taraxacum officinale)* are familiar to anyone who has ever tried to maintain a lawn. The cheerful flowers open each morning like gold coins on green velvet. If that was all they did, they would probably be welcome lawn plants. However, they close by evening and begin rapidly elongating their stems, presenting their fluffy seed heads to be blown away by the wind. A freshly mowed lawn can be full of these unsightly intrusions in a day's time. More effort and money have probably been spent in attempts to rid lawns of Dandelions than any other lawn pest. When our children were small, we used to pay them by the well-packed grocery bag to dig Dandelions from the lawn. Instead of diminishing, the Dandelion density increased from year to year. We decided that the roots left in the soil after the digging must be sending up multiple shoots. About the only effective ways to eliminate Dandelions are to use a chemical herbicide or else let the grass grow tall enough to crowd them out.

Winter Cress or Yellow Rocket *(Barbarea vulgaris)* is a common yellow-flowered plant growing in fields and waste areas. The one to three foot branched stalk bears terminal clusters of tiny, four-petaled flowers and broad, glossy, fan-shaped stem leaves that are deeply cut. The lower leaves have a large end lobe and two to four pairs of rounded side lobes.

The small yellow flowers at upper right are those of Golden Alexanders *(Zizia aurea)*. The tiny flowers are borne in clusters of small umbels, each widely separated from those around it. The stems for the flower umbels in a cluster all radiate from a common point. A perennial growing one to two feet tall, Golden Alexanders is a member of the Parsley family.

Golden Corydalis *(Corydalis aurea)* grows about a foot tall and has bright green, much-divided fern-like leaves. Groups of bright yellow half inch flowers bloom at the tips of leafy stems. The flowers have a two-lipped appearance, but are composed of four petals—two large outer ones and two small inner ones. Corydalis, a member of the Poppy family, is related to Dutchman's Breeches (Plate 6) and the other Dicentras.

Star of Bethlehem *(Ornithogalum umbellatum)* is a garden flower that has become naturalized in many areas and is often found in abundance around old homesites. It grows from a bulb and has numerous grassy leaves that have a whitish stripe down the center. The waxy white, six-petaled flowers are about one and one-fourth inches wide and have a green stripe on the outside of each petal.

At the bottom of the painting is Myrtle or Periwinkle *(Vinca minor)*. This beautiful evergreen trailing vine is often used as a groundcover under trees. One inch blue flowers bloom in the axils of dark green, shining leaves. Vinca forms a dense carpet of foliage that, in shaded areas, prevents the growth of other plants. In some parts of the state, Vinca has naturalized over large areas and become a problem by crowding out native woodland wildflowers.

Inspecting the Vinca blossoms is a Great Spangled Fritillary.

April 12, 1986

Plate 9

The specimens for Plate 9 were found near Lake Monroe in Monroe County. On the left is Creeping Phlox *(Phlox bifida)*. Other common names are Sand Phlox and Cleft Phlox. Sand Phlox refers to this plant's tolerance for dry sandy soils, and Cleft Phlox alludes to the deeply indented shape of the petals. Creeping Phlox grows only three or four inches tall. The creeping stems will root readily at nodes where the upright flower heads form. The one-half to three-fourths inch flowers are pale purple and their deeply notched petals join to form a slender tube. The pointed stiff, narrow leaves grow as long as one and one-half inches.

Confederate Violet *(Viola papilionacea var. priceana)* is shown in the center of the painting. This violet is generally considered to be a form of the common blue violet. The lower petals of the large, grayish-white flower are marked near the center with lavender areas and dark veins. Large, heart-shaped leaves have scalloped margins and prominent vein structure. Leaves and flowers rise from the crown of the plant on separate stalks. The flowers have stems long enough to allow picking for small bouquets, and the young leaves are high in vitamins A and C and can be used in salads or cooked as greens. Purple Cress *(Cardamine douglassii)* is shown on the right. Racemes of rose-purple or pink flowers rise from basal rosettes of roundish, long stalked leaves. The leaves on the flower stalk are elongated and without stalks. The four-petaled flowers are one-half to three-fourths inches across and grow on stems six to twelve inches tall.

The combination of plants in this painting reflects the tolerance that many wild flowers possess for habitats that deviate from ideal conditions. Although specimens for this painting were found growing quite close together, Creeping Phlox characteristically prefers dry soil or well-drained banks. In some areas of southern Indiana, it makes a haze of lavender over the rocky slopes along highways and country roads. Purple Cress is characteristically found in moist soils of woods or valley floors. Confederate Violets enjoy the same conditions as Common Blue Violet and may be found along the edges of woods or in meadows. They are vigorous enough to invade lawns where the soil has been allowed to become acid, tolerating fairly close mowing and growing happily among the lawn grasses.

The tree seedling is an American Elm in its second or third year of growth. This once-abundant species of large, graceful trees has been ravaged by Dutch Elm Disease, a fungus accidentally introduced into this country in the 1930s and spread by elm bark beetles. Young trees still continue to sprout and reseed themselves, although they no longer achieve the grandeur their longevity once permitted.

April 12, 1984

Plate 10

Field Pansy *(Viola kitaibeliana),* at lower left in Plate 10, is an annual plant of European origin that grows three to eight inches tall. The one-half inch five-petaled flowers are whitish or pale lavender blue in color. Both flower stems and large lobed, blunt-tipped stipules arise from the leaf axils. The spoon-shaped leaves are small.

Above the Field Pansies is Henbit *(Lamium amplexicaule).* The lavender flowers of this small plant are found in whorls at the axils of the horizontally held leaves. The one-half to two-thirds inch hairy flowers have two lips with four stamens that protrude beyond the face of the flower. The opposite, stalkless leaves have a rounded shape with scalloped edges and are one-half to one and one-half inches wide.

At the left side of the painting is Smooth Rock Cress *(Arabis laevigata).* This plant is slim and sparse in appearance. The thin, lance-shaped leaves are pale gray-green in color, and the tiny flowers are greenish-white. The plant produces numerous long, thin seedpods that either curve downward or extend horizontally from the stem.

Pendulous two inch blossoms of Wild Columbine *(Aquilegia canadensis)* hang from slender stems at the top of the painting. This curiously formed flower is composed of five sepals and five petals. The latter are yellow at their tips and extend upward to form long, red spurs. Between each petal is a red sepal. Numerous yellow stamens hang gracefully below the flower. The tips of the spurs contain nectar that attracts the long-tongued insects needed to pollinate the flower. The long-stalked leaves are compound and may have from nine to twenty-seven light green three-lobed leaflets. A short-lived perennial, Wild Columbine reseeds itself generously. It thrives on rocky bluffs or well-drained hillsides and makes a choice addition to the flower garden.

The bell-shaped blue flowers below the Wild Columbine are those of Greek Valerian *(Polemonium reptans).* This is a smooth, weak-stemmed plant that forms low mounds of long, pinnately compound leaves. Leaflets are ovate to lanceolate and about one and one-half inches long. The one-half inch, five-lobed corollas are borne in loose clusters. A three-lobed stigma extends beyond the tips of the corolla lobes.

At the bottom of the painting is Corn Speedwell *(Veronica arvensis),* a low, hairy, much-branched plant bearing minute, four-petaled violet-blue flowers in the upper leaf axils. Paired lower leaves are about one-half inch wide and are round in shape. Upper leaves are stemless, alternate, and lance-like in shape. This plant grows two to six inches tall and is often invasive in lawns and gardens.

We found these specimens at Harmonie State Park.

April 14, 1986

Plate 11

The specimens for Plate 11 were discovered at McCormick's Creek State Park along the limestone ledges overlooking the creek. In this area McCormick's Creek has cut a deep, narrow canyon through the layers of bedrock and provided a setting of great beauty for a rich variety of wildflowers. The fertile pockets of organic soil and forest litter, the gentle seeping of cool water from crevices and bedding planes in the rock strata, and the protection from wind and sun provided by the canyon walls produce a secure habitat for the lush growth of delicate wildflowers that abound in this magic place.

For Maryrose and me, walking in this canyon was like getting a hug from Mother Nature. There were no vistas because of the limestone cliffs and the bends in the stream, and yet there was no feeling of claustrophobia. Fallen blocks of rock supplied places to sit and rest, and the only sounds besides our own were the happy tricklings of the nearby stream. We felt protected and secure, just like the wildflowers growing there.

The tall plant at the left of the painting is Shooting Star *(Dodecatheon meadia)*. The smooth, hollow stem holds an umbel of rose, lilac, or white flowers six to twenty inches above a basal rosette of light green, oval leaves. Frequently leaf bases and sometimes leaf midribs are red. The flowers are one to one and a half inches long and have five strongly backward pointing petals. Five yellow stamens closely surround the single pistil in a cone-shaped arrangement at the tip of the flower. A narrow dark band of purple separates the yellow of the stamens from the base of the petals. This unusual shape suggests the common name Shooting Star. Other names related to shape are Mosquito Bills and Birdbills. In the days of the settlers, a common name was Prairie Pointers because it was then found more frequently in prairie or meadow habitat than it is today. It is also called American Cowslip because of its relationship and similarity to the English Primrose or Cowslip.

The two specimens at the top of the painting are male and female forms of the same plant, Meadow Rue *(Thalictrum dioicum)*. The female plant is on the left. The tiny up-facing flowers consist of pistils protruding from small, cupped sepals. The male plant on the right has pendulous flowers with prominent yellow or greenish stamens hanging downward from the cupped sepals. Neither flower has any petals. The primary beauty of the plant, which grows one to two feet tall, lies in its numerous dainty three-lobed leaflets, reminiscent of Maidenhair Fern.

At the lower right are Plantain-leaved Pussytoes or Early Everlastings *(Antennaria plantaginifolia)*. Their one and one-half to three inch oval leaves have three or more veins and form a basal rosette. Stem leaves are pointed and decrease in size toward the top of the stem. The large, dark leaves are from the previous season. This plant spreads by means of stolons or runners, and can form large, dense mats of basal rosettes. When in bloom the fuzzy flowers can suggest a fog-like haze over the ground.

Like Meadow Rue, this plant also has male and female flowers on different plants. When the flowers are mature, stamens bearing brown specks of pollen extend beyond the white fuzz in the male flower, and Y-shaped pistils extend beyond the fuzz in the female flower. Female flowers frequently have a pink tint. The flowers in the painting are not mature enough to show either the stamens or pistils. There are many species of Pussytoes that are similar in appearance.

The plant in the foreground is Wild Ginger *(Asarum canadense)*. Both the heart-shaped or kidney-shaped leaves and the leaf stems are fuzzy. The leaves grow in pairs, each on its own stem, arising from an underground runner. The purplish brown flowers with their cup-like calyx bloom at ground level between the pair of leaf stems. The underground rootstock has a spicy, ginger-like aroma when crushed; the pioneers used it as a substitute for expensive imported ginger. It was also used medicinally as a poultice for cuts and sores and has been found to contain a natural antibiotic. Wild Ginger grows only four to six inches in height at bloom time. As the season progresses, however, the leaves enlarge considerably and grow taller.

Bluets, Innocence, or Quaker Ladies *(Houstonia caerulea)* are shown at the base of the Shooting Star. These ethereal little flowers grow only two to six inches tall and are usually blue or lavender, though they can be white or pink. The four petals join at the center to form a yellowish white tube. The slender flower stems arise from a basal rosette of tiny leaves. There is frequently a pair of opposite leaves halfway up the stem. Though tiny and delicate, this plant can appear in such numbers that it produces drifts of color.

April 15, 1985

Maryrose Wampler

Plate 12

In an upland area at Spring Mill State Park where a thin layer of soil covers the limestone bed-rock, we found the specimens for Plate 12. All of these plants need sharp drainage and can tolerate dry soil conditions.

Two color forms are shown of Birdfoot Violet *(Viola pedata),* the lavender form in the foreground and the less common bicolored form at the top of the painting. The flowers of this species are the largest and showiest of the genus Viola, sometimes reaching as much as one and one-half inches in width. The broad, richly colored petals are further enhanced by orange stamens visible in the flower's throat. The bottom petal in both color forms has purple veining on a light background. Leaves are palmately divided into irregularly lobed leaflets. Flowers are held well above the leaves and are borne on separate stems four to ten inches tall. This flower, particularly the bicolored form, is considered by many wildflower lovers to be the most beautiful violet in the world, and a grassy hillside dotted with heavily blooming clumps of them is indeed a breathtaking sight.

The yellow daisy-like flowers at the top of the painting are Round-leaved Ragwort *(Senecio obovatus),* whose three-fourths inch flowers are borne in a flat-topped cluster on a six to twenty-four inch stem. Basal leaves are broadest above the middle and have blunt tips. They taper to the base of the leaf stem and are sometimes lobed. The stem leaves are usually pinnately lobed. Round-leaved Ragwort prefers drier growing conditions than its near relative Golden Ragwort (Plate 21).

The small blue flowers are those of Blue-eyed Grass *(Sisyrinchium montanum).* A tiny iris, Blue-eyed Grass has half inch blue flowers composed of three petals and three petal-like sepals, each tipped with a thorn-like point. This species is characterized by an unbranched flower stem with no bract midway up. At the top is a cluster of flowers and buds with a pointed bract extending above them. The grass-like leaves are four to twenty inches long and about one-fourth inch wide.

The fleshy little plant at lower right is Pennywort *(Obolaria virginica),* a spring blooming member of the Gentian family. This three to six inch plant has half inch dull white or pale purple four-petaled flowers borne in clusters in the axils of purplish, wedge-shaped bracts. The paired leaves below the flowers are round in shape, while the leaves lower on the stem are scale-like.

In the foreground is Stargrass *(Hypoxis hirsuta).* A diminutive member of the Amaryllis family, Stargrass bears small clusters of yellow one-half to three-fourths inch flowers at the tops of slim, leafless three to six inch stalks. Arising from a tiny bulb, hairy, grass-like leaves grow four to twelve inches long. Stargrass begins blooming in the spring and may bloom intermittently throughout the summer.

Hoary Puccoon *(Lithospermum canescens)* has luminous golden, tubular flowers with five flaring lobes. Numerous one-half inch flowers are clustered at the tops of erect six to eighteen inch stems. Narrow, alternate leaves are one and one-half to two inches long and are covered with short, silvery hairs. Puccoon is an Indian word meaning a plant that yields yellow dye.

The area that produced these specimens is an example of the unpredictability of nature. Surrounded by rich, moist, deciduous woods, it is a little slice of pure prairie. An astonishing number of species grow there that should not be found within a hundred miles or more. Whether it simply happens to offer favorable growing conditions or is a remnant of a past age when these plants were common in the area poses an interesting question. Since this land has not always been part of the park system, it is conceivable that it could even be an old wildflower garden planted by a knowledgeable and well-traveled plant fancier.

April 17, 1986

Maryrose Wampler

Plate 13

We came upon the specimens for Plate 13 in a wooded valley bordering the stream at McCormick's Creek State Park. The painting shows Dwarf or Spring Larkspur *(Delphinium tricorne)* at lower left. With their distinctive cone-shaped cap of flowers, these interesting plants look a bit like elves standing sentinel along forest passages. There are often several plants in the same vicinity, but they tend to keep "space" around themselves so that they are seen as individuals. The three-fourths inch flowers are most commonly purple, but may be white, lavender, or a combination of those colors. An individual floret consists of five large colored sepals and four inconspicuous petals at the center. The upper sepal is extended about one-half inch to form an erect spur. Flowers bloom first at the bottom of the stem. The upward curving spurs are prominent even on small unopened buds and give the stem of flowers a lilting look. Most of the buttercup-like leaves arise from the base of the plant. Those on the stem are alternate and become progressively smaller.

The common name, Larkspur, is derived from a similarity in appearance of the flower's spur to the spur on a lark's foot. The Latin species name, tricorne, refers to the three-horned seedpod that develops from each pollinated flower. Larkspurs contain a harmful alkaloid that is poisonous to grazing cattle.

Above the Larkspur is Drooping or Bent Trillium *(Trillium flexipes).* The large, showy white flowers may be as much as one and three-fourths inches across and are carried at an angle to the main stem, sometimes drooping beneath the large, broad trio of leaves. This lovely flower is common along wooded ravines and valley floors where the soil is deep and moist. It blooms a week or two after the maroon Prairie Trillium. For many years I mistakenly called this flower Nodding Trillium *(Trillium cernuum)* because the name aptly describes the appearance of the flower. The creamy-white color of the anthers, however, distinguishes this flower as Trillium flexipes. T. cernuum has pink anthers. Apparently this characteristic remains constant even though the color of the flower itself may occasionally be maroon instead of white. Nodding white trilliums found in most of Indiana are T. flexipes. T. cernuum is probably a more common trillium nationally but is found in more northern areas or in regions of higher elevation.

Virginia Waterleaf *(Hydrophyllum virginianum)* is also a common denizen of rich, moist valley floors. The bell-shaped flowers are lavender to white in color and are composed of five joined petals. The flowers are carried above the foliage on long stalks that arise from the leaf axils. Hairy filaments extend beyond the petals and give the flower clusters a fuzzy appearance. The smooth five to seven lobed leaves are triangular in their general outline and sometimes have a mottled look as though stained with water—hence the name Waterleaf. Growing one to three feet tall, this plant often forms colonies that provide drifts of lavender color in the woodland spring landscape.

Smooth Yellow Violet *(Viola pensylvanica)* is characterized by its smooth stem, the one to five basal leaves, and the untoothed stipules at stem junctures. Upright, heart-shaped leaves sometimes hide the yellow flowers. The flowers grow on the same stem as the leaves and have purple stripes on the lower petals.

The mushroom is probably *Morchella semilibera,* an early morel that is popularly known as a Blackhead in southern Indiana. Although small, it possesses the famous morel flavor and is eagerly sought after in the spring. Plate 13 was painted in a poor morel year, and Maryrose's hopes of morel omelet cooked over the campstove were doomed to failure. This diminutive specimen, found by a visiting brother-in-law, comprised the entire crop for the season.

April 18, 1985

Plate 14

The specimens for Plate 14 can be seen at Spring Mill State Park. At lower left is an interesting striped violet that, except for the color, appears to be *Viola papilionacea* (Plate 16). We used to think that the broken color was produced by a virus similar to that which causes the color breaks in Rembrandt tulips. It is a phenomenon we have observed at several sites in southern Indiana. Several years ago we moved a similar plant to our garden from the Indian Springs area. We noticed that the broken color pattern did not spread to other violets in our garden; moreover, it became progressively less marked each year. It is possible that the striping is related to special characteristics of the soil at sites where these violets are found.

Standing above the striped violet is the purple form of Jack-in-the-pulpit *(Aresaema triphyllum)*. There is a green form of this plant (Plate 15) that is generally accepted as being the same species. Because of its unusual shape and descriptive name, this is a wildflower that most people remember once they have made its acquaintance. These plants grow from corms that may produce larger plants as they age, but do not generally multiply vegetatively. New plants are produced from seeds contained in the large head of handsome red berries that ripens in the fall. Large, mature specimens of Jack-in-the-pulpit are usually solitary; or, if there are numerous plants in an area, they have space between them and are not clumped. Possibly the "Jacks" or preachers in these pulpits do not care to hear too clearly the opinions of their neighbors.

At the top of the painting are the six-pointed flowers of Blue Cohosh *(Caulophyllum thalictroides)*. These flowers are composed of six yellow-green to purplish petal-like sepals and six inconspicuous petals. They are replaced later in the season by large, beautiful but poisonous blue berries. As suggested by the species name, the foliage of this attractive gray-green plant resembles that of Meadow Rue *(Thalictrum dioicum),* shown in Plate 11. While the one to three foot plant has a delicate, leafy look, it actually has only two true leaves. A large compound leaf about midway up the stem is subdivided into usually twenty-seven three-lobed leaflets. A smaller compound leaf is found just below the cluster of flowers.

Below the Cohosh is the white form of Dwarf Larkspur *(Delphinium tricorne)*. The more common purple form of this flower is shown in Plate 13. The white pyramidal clusters of flowers with their lavender or blue shadings are prominent against the dark greens and browns of forest surroundings. Dappled patterns of light and shadow, continually shifting as sunlight filters through the canopy of tree leaves, give this plant a ghostly beauty.

In the foreground is Three-lobed Violet *(Viola triloba),* a violet that can be tricky to identify because the shape of the leaves it produces varies with its period of growth. The earliest leaves are smooth, un-lobed, and rounded in shape. Later, as the plant blooms, the leaves produced are softly hairy and have three lobes of varying depth. The flowers are blue and have bearded areas at the throat on the lower three petals. Since the flowers of this woodland violet are similar to those of several other species, identification is based upon the shape of the later, principal leaves.

April 19, 1986

Plate 15

We discovered the specimens for Plate 15 growing along the stream at McCormick's Creek State Park. The yellow flower at left foreground is Buttercup *(Ranunculus hispidus)*. The Latin name means "little frog," presumably because it blooms where frogs live. The glossy yellow flowers are about three-fourths of an inch wide and consist of five petals and five sepals. The leaves are divided into three segments and begin in the spring as a basal rosette about six inches tall. As the season progresses, the stems elongate and become runner-like, growing to twenty-four inches. Buttercup foliage is considered poisonous to livestock, but that seldom becomes a problem because cows do not like its bitter taste.

Bellwort *(Uvularia perfoliata)* is named after the uvula in the throat and was once thought to be a cure for disease in that location. The attractive yellow flowers with their curiously twisted petals hang in pairs from an arching six to eight inch stem that forks at the tip. Orange granules line the inside of the flower petals. The smooth, parallel veined leaves have a whitish bloom and appear to be pierced by the stem.

White Baneberry *(Actaea pachypoda)* has a single oblong cluster of tiny white flowers consisting of four to ten very narrow petals and long bushy stamens. Divided and subdivided, the dark green leaves provide a handsome setting for the prominent cluster of berries that develop later in the season. The berries, commonly called "doll's eyes," are white with a black dot in the center. They grow on thick, bright red stalks. Both the berries and the coarse fibrous rootstock are poisonous.

This painting also shows the green form of Jack-in-the-pulpit *(Arisaema triphyllum)*. The purple form, shown in Plate 14, is generally accepted as a variation of the same species. The "flower" consists of a flap-like spathe, often striped, enclosing a club-shaped spadix (the "Jack," or preacher, in his canopied pulpit). The true flowers are found at the base of the spadix. One or two long stalked leaf stems are topped by a three lobed leaf.

Male and female flowers are often found on different plants. If the flowers look like minute green berries, the plant is female; if they look like threads, the plant is male. Jack-in-the-pulpit is quite variable as to sexual orientation. Sometimes both male and female flowers are on the same plant, or a plant may produce male flowers one year and female flowers the next. This unusual behavior appears to be controlled by nutrition. With a rich food supply the plant can produce the bright cluster of berries that contains the seeds. If the plant is not "feeling robust," it produces only male flowers and rests for a season.

Jack-in-the-pulpit is also called Indian Turnip because it was used by the Indians as a source of food. The corm from which the plant grows contains calcium oxalate crystals that can severely burn the mouth and throat. But the Indians knew that if they boiled the corms to remove the calcium oxalate, they could then dry and grind them into flour to be made into cakes.

The white flower in the right foreground is False Rue Anemone *(Isopyrum biternatum)*. The five petal-like sepals form flowers one-half to three-fourths inch wide. The deeply lobed leaflets are in threes. Growing four to ten inches tall, False Rue Anemone forms large colonies of leafy plants. Its lush growth habit and the deeply lobed leaves differentiate it from the more delicate true anemone.

April 20, 1985

Plate 16

It was along a small tributary valley of Brummett's Creek in Monroe County that these specimens were found. The small white flower at lower left is Star Chickweed *(Stellaria pubera)*. The pointed, dark green leaves are rounded at the base and most are attached to the stem without a leaf stalk. Five broad green sepals are a bit shorter than the petals and emphasize their whiteness. The five white petals that comprise the one-half inch flower are notched so deeply that there appear to be ten petals instead of five. Though related to the hated weed that chokes our gardens, this charming woodland flower has neither its invasive growth habit nor its tendency to spread rapidly. In fact, when given a shady spot in our flower garden Star Chickweed behaved impeccably, making a neat hemispherical mound of foliage covered with white star-like flowers each spring.

Wild Geranium *(Geranium maculatum)* is a conspicuous five-petaled flower one to one and one-half inches wide. The broad veined petals show many shades of rose-lavender and vary considerably in color intensity. The deeply cleft leaves, two to six inches wide, consist of three to seven lobes. Basal leaves are largest and have long stalks. Smaller stem leaves are opposite and have fewer lobes and shorter stalks. The reason for one of its common names, Cranesbill, becomes apparent as the seed pod begins to develop: the center of the flower elongates into a beak-shaped capsule containing five seeds. This plant has a long season of bloom in the spring. Since the flowers are large and held above the foliage on one to two foot plants, Wild Geranium sometimes produces a spectacular display. Many times in our search for interesting wildflowers we have stopped our car, backed up, and gotten out to check out a gorgeous drift of lavender color—only to find the familiar Wild Geranium.

Virginia Bluebells *(Mertensia virginica)* is also commonly called Virginia Cowslip. Drooping racemes of bell or trumpet-like flowers of soft blue open from dusty pink buds. Occasionally a plant may be found whose flowers remain pink after opening. These distinctively colored three-fourths to one inch flowers are borne on one to two foot plants. The large oval opposite leaves are strongly veined. The flower produces a seed capsule composed of five compartments, each of which contains one seed. Although a perennial, it spreads over broad areas by means of seed when it finds a suitable habitat. It is not uncommon to find an entire floodplain carpeted with these showy flowers. Such an expanse of pure blue color is awe-inspiring. These plants have a short season of growth; they have totally disappeared by the end of spring.

While the common name for *Trillium recurvatum* is Prairie Trillium, it is probably the most common woodland Trillium in many parts of the state. The plant grows six to eighteen inches tall and has three erect lance-shaped petals of various shades of reddish or brownish maroon. The three green sepals curving downward from the base of the petals provide the species name. The three leaves have short petioles and are often strongly mottled with green or brown.

Blue Phlox *(Phlox divaricata)* is often called Wild Sweet William. That is a misnomer, because the real Sweet William is a Dianthus—a member of the Carnation and Pink tribe—rather than a Phlox. This plant grows nine to eighteen inches tall and has terminal clusters of one inch flowers of bluish violet. Five wedge-shaped petals join to form a tube. Petal shape may be smoothly rounded or it may be indented or "cleft" to varying degrees. Slender stems may be hairy or sticky toward the top and bear opposite leaves that are oval or lance-shaped. The leaves are tough and leathery and may persist through the winter. Blue Phlox is often a prominent flower in the spring woods or on wooded banks along roadsides. Occasionally a pure white form is found. While looking for specimens for Maryrose to paint, we came upon a wooded plot that was carpeted exclusively with "Blue" Phlox of pristine whiteness.

Common Blue Violet *(Viola papilionacea)* is also called Meadow Violet because it is so common and numerous in meadows. If acid conditions are present, it may invade lawns. This flower is dark violet in color and is held above the foliage on its own stem. Two lateral petals are bearded and the three lower petals are veined. The bottom petal is longer and forms a spur that extends behind the flower. In addition to the normal flowers, there are flowers on short stems near the ground that fail to open, but that produce enormous quantities of seed. The large strongly veined leaves are heart-shaped and have scalloped margins. Each leaf is borne on a long petiole that arises from the rootstock. Violet leaves can be cooked as spring greens or used in salads. The flowers are sometimes made into candies or jellies.

April 25, 1984

Plate 17

Specimens for Plate 17 were found on private land near North Vernon, Indiana. White or Pale Violet *(Viola striata)* is shown at the left of the painting. This violet grows six to twelve inches tall and is characterized by large, toothed stipules at the base of the heart-shaped leaves. Later in the growing season when the plant is taller, this characteristic is more apparent. The cream-colored to milk-white flowers, each with its purple-striped lower petal, grow on the same stalk as the leaves. This lovely violet frequents low woods and stream banks.

In the center of the painting, Crested Dwarf Iris *(Iris cristata)* is being inspected by a Spring Azure butterfly. The flower's name comes from a crest or pleated projection on each down-curving sepal. In my experience, people more frequently call this flower by its scientific name than by its "common" name, and often they simply say "cristata."

Iris cristata is a wildflower of unusual beauty. The two and one-half inch lavender flowers are held above the foliage and cover colonies of plants that may be several feet across with masses of color. As the flowers mature, leaves elongate and mask the faded flowers. At bloom time, the flowers are only four to five inches tall. Later in the season, the strap-shaped leaves may reach one foot in length.

This wildflower is frequently sold through nursery catalogs as a groundcover, although it is hardly aggressive enough to merit that designation. There is a white form and also a darker color variation. Iris cristata is found in nature in well-drained, organic woodland soils in shaded locations. When grown in the flower garden it can tolerate fairly bright sunlight, but needs to be protected from taller growing plants that encroach and crowd it out.

The textured leaves in the center and also in the right foreground of the painting are Dwarf Ginseng *(Panax trifolium)*. The plant in the foreground shows the round umbel of tiny white flowers, and the specimen in the background shows the early growth stage of the yellow berries that bear the seeds. Below the flower head is a whorl of three compound leaves, each composed of three to five stalkless, toothed leaflets. The round tuber-like root is edible and can be used either raw or boiled. Dwarf Ginseng grows only four to six inches tall. While not subject to the high commercial demand of its larger cousin, *P. quinquefolium,* Dwarf Ginseng is uncommon in Indiana.

A specimen of Wild Stonecrop *(Sedum ternatum)* is shown at the right. Its flower spray usually consists of three arched, horizontal branches lined on the upper side with white five-pointed flowers. The smooth, fleshy leaves are in whorls of three. The succulent stems lie along the ground and send up vertical bloom stalks that grow four to eight inches tall. Stonecrop is another wildflower that is sometimes used in the garden as a groundcover in moist shaded areas.

The butterfly shown in this painting, and also in Plate 11, is a Spring Azure. Emerging from overwintering pupae, these are the earliest of the native butterflies to emerge in spring—sometimes so early that snow is still on the ground. Spring Azures search out the first flowers that bloom.

April 28, 1984

Plate 18

Two-flowered Cynthia or Dwarf Dandelion *(Krigia biflora)* is a common flower along country roads and woodland paths. A rosette of blue-green leaves forms the base of the plant. Both smooth and coarsely toothed leaves may be found together. Small upper leaves clasp the stem and occur only where the stem branches. The flowers look like those of the Dandelion but are only about half as big, and tend to be oranger in color. Growing to a height of two feet, Two-flowered Cynthia has a slim, delicate appearance. The stems and leaves blend with their surroundings in such a manner that the flowers can give the impression of being suspended in space a foot or two above a grassy bank.

To the right of the Two-flowered Cynthia is Large (or Lily-leaved) Twayblade *(Liparis lilifolia)*. A single bloom stalk four to ten inches tall arises from two broad, oval, glistening leaves bearing a loose raceme of anywhere from five to forty half inch flowers. The thread-like appearance of its pedicels and flower parts and the translucence of its broad mauve or purplish lip make it a flower of unusual delicacy. This diminutive member of the orchid family is widely distributed and fairly common, but seldom occurs in large numbers. It likes mossy woods or pine forests and is often seen as a single plant or a small grouping of fewer than a half dozen plants growing together at the base of an oak or pine tree. The generic name, Liparis, means "fat" or "shining," in reference to the greasy shine on the leaves. These leaves are often seen growing in isolation among brown pine needles or dead leaves of the forest floor. The plant's tolerance of acid soil and its low light requirements enable it to thrive under conditions that most vegetation would find inhospitable.

Paired yellow-green bell-shaped flowers dangle from the axils of the alternate sessile leaves of Solomon's Seal *(Polygonatum biflorum)*. Later in the season blue-black berries about one-fourth inch in diameter will hang beneath the leaves of the arching stem. The common name comes from the circular scar or seal left on the upper surface of the rhizome where stalks from previous years were attached. Solomon's Seal has been used as a source of food by both Indians and pioneers. The tender shoots can be eaten in the spring like asparagus, and the rhizomes can be cooked as a starchy vegetable.

Alumroot *(Heuchera americana)* is related to the cultivated garden flower Coralbells. The tiny flowers droop from short branching stems and are usually greenish or green tinged reddish. The prominent stamens extend well beyond the petals and are brightly tipped with orange anthers. The leafless flower stalk rises two to three feet above a rosette of maple-shaped basal leaves. Alumroot has been used as an astringent to relieve skin disorders and to treat diarrhea.

Violet Wood Sorrel *(Oxalis violacea)* is a small violet to rose-purple, five petaled flower about one-half to three-fourths inch across. The inversely heart-shaped leaves look like clover leaves. All leaves are basal and grow from a scaly, bulb-like corm that sends out runners to form dense colonies. The leaves are frequently reddish or purple beneath and may have a purple pattern on top.

April 28, 1985

Plate 19

Specimens for Plate 19 were encountered at Tippecanoe River State Park. Wild Strawberry *(Fragaria virginiana)* is a hairy plant that grows three to six inches tall. Each slender leaf stalk arises from the crown of the plant and bears at its top three coarsely toothed leaflets one to one and a half inches long. The flowers are borne in flat clusters on separate stems that tend to be shorter than the leaf stems. The three-fourths inch flowers have five rounded white petals and numerous stamens and pistils that cover a dome-like central structure which becomes the berry as the seeds mature. There are five pointed green sepals that persist to form a cap for the ripe strawberry. In addition to propagating itself by means of seeds, Wild Strawberries produce numerous runners. The plants often cover large areas in thin woods or open meadows with dense mats of foliage. Cultivated strawberries involve F. virginiana in their parentage, but have been bred to produce larger and firmer fruit. Wild Strawberries are only about one-half inch in diameter, and so soft that they become mushy shortly after picking. If they can be used quickly, however, their intense, penetrating flavor is matchless. Later in the season, we saw this plant growing in profusion among the pines on a Christmas tree farm. Maryrose had been having trouble convincing a milkweed she was trying to draw to hold its florets in a natural position, so we had driven our truck camper out into the middle of the field of trees to set up next to a living specimen. Strawberries were everywhere, so ripe that their intoxicating perfume made it difficult to think of anything else. Knowing they were there, we had come prepared; I abandoned my usual role as a research assistant and produced eight pints of wild strawberry freezer jam while Maryrose drew the milkweed.

The tiny yellow flowers above the strawberry plants are those of Wormseed Mustard *(Erysimum cheiranthoides)*. The four-petaled flowers are borne in terminal clusters on plants six to thirty-six inches tall. Leaves are one to four inches long and are sometimes slightly toothed. Seedpods stand erect and are long and slim.

Northern Downy Violet *(Viola fimbriatula)* is a small, hairy plant. Arrow-shaped or egg-shaped leaves are on short stalks. The violet-colored flowers are large in proportion to the plant and are held well above the foliage on separate stems.

Swamp Buttercup *(Ranunculus septentrionalis)* has showy, bright yellow flowers that are so glossy that they appear to have been varnished. The one-inch flowers have five rounded petals and five narrow green sepals surrounding a center containing numerous stamens and pistils. The three wedge-shaped segments which comprise the leaves are short-stalked and may be one and one-half to four inches long. The plant is compact early in its bloom season, but quickly elongates its reclining stem to as much as three feet. It later forms roots at each leaf node.

The normal habitat for Swamp Buttercup is much wetter than that required by the Wild Strawberry. This specimen was found on alluvial soil that was sandy in texture and well enough drained that we felt it could be combined with the strawberry and violet. It wasn't until after the preliminary work on the painting was done that we discovered the "swamp" part of the common name.

The butterfly is a Silvery Checkerspot.

April 28, 1986

Maryrose Wampler

Plate 20

The wildflowers shown in Plate 20 were found along a woodland trail at Pokagon State Park. At lower left is Downy Yellow Violet *(Viola pubescens)*. This striking flower often has one basal leaf with stem leaves and flowers on the same five to sixteen inch stalk. Young leaves and upper stem are downy. The five-petaled yellow flower may be as large as three-fourths inch across. The lower petal is veined with purple and the two lateral petals are bearded. The toothed leaves are two to five inches long.

Showy Trillium *(Trillium grandiflorum)* has the largest flower of any Trillium. The three broad, wavy white petals form a flower that may be as much as four inches across. As the flower ages, it usually turns pinkish. The center flower in the painting shows the beginnings of this color change. After the flower withers, a six-lobed black fruit develops. The fruit contains seeds, each of which has a light-colored strophiole (a kind of crest) that attracts ants. The ants carry the seeds to their nest, eat the strophioles, and discard the seeds, thus establishing new plantings of Showy Trilliums. These flowers are large and impressive, but their beauty should be enjoyed only in the woods. Since their three leaves are all at the top of the stem, picking them prevents renewal of their food supply and causes the death of a plant that has taken nature several years to mature to blooming size.

The small, frothy flower at upper left is Goldenseal or Yellowroot *(Hydrastis canadensis)*. This plant tends to grow in colonies. A plant unit is composed of a single basal leaf on a long petiole and a bloom stalk bearing two alternate stem-leaves. The flower is just above the second stem-leaf so that when the red raspberry-like fruit develops, it appears to be nestled at the base of the upper leaf (see Plate 50 for Goldenseal in fruit). The flower consists of a puff of white stamens surrounding several pistils. It has no petals, and the sepals drop as the flower opens. The veined, palmately-lobed leaf continues to grow after flowering time, eventually becoming six to eight inches across.

A colony of Yellowroot is more beautiful in fruit than in flower. The deep green leaves, jeweled with large bright red berries, may grow densely enough to obscure the ground beneath them. Their large yellow rootstocks were used medicinally by both Indians and pioneers as a tonic, stimulant, and astringent, but primarily for stomach ailments and as a laxative. The yellow pigment in the root is suitable for use as a dye.

At the top of the picture is Mayapple or Mandrake *(Podophyllum peltatum)*. This plant produces a single white flower one and a half to two inches across, blooming in obscurity beneath two large umbrella-like, palmate leaves. Springing from underground stems, Mayapple often forms large colonies of one to one and a half foot plants. Specimens that are not old or strong enough to bloom have only one leaf. Although the plant parts and green fruits are poisonous, the yellow, ripe, two-inch fruits are edible. Mayapple's caustic resin causes diarrhea and vomiting, and has been used externally to treat warts and tumors.

The plant at lower right is Wood-betony or Lousewort *(Pedicularis canadensis)*. The low hairy plant grows six to fifteen inches tall and is topped by a dense whorl of tubular, hooded flowers that are reddish, yellow, or both. Many of the fuzzy fern-like leaves form a basal rosette. Stem leaves tend to be smaller. The common name Lousewort comes from an old belief that cattle and sheep that graze on Betony are likely to have lice. Although interesting and attractive, Betony is difficult to transplant because of a symbiotic dependence on a certain soil fungus associated with its roots.

At the bottom of the painting is Spurred Violet *(Viola rostrata)*. The common name refers to the half inch spur on the lower petal. This distinctive violet is pale lavender in color and has dark lines on the three lower petals.

April 29, 1985

Plate 21

One of the bog areas at Pokagon State Park supplied the specimens for this painting. While only the Marsh-marigold is truly aquatic, the other plants need or at least tolerate wet ground.

The white flower on the left is Spring Cress *(Cardamine bulbosa)*. The basal leaves are roundish in shape and are long stemmed, while the stem leaves tend to be sessile. In seepage areas or wet fields this plant often forms large, dense colonies that create drifts of pristine white against the fresh spring green of surrounding plants. The bulb-like root has a mild, horseradish-like flavor and was used as food by both Indians and pioneers.

At the top of the painting is Golden Ragwort *(Senecio aureus)*, also called Groundsel or Squaw-weed. Golden Ragwort is frequently seen in fields that have not been plowed for a year or two. It is so rank in growth that such fields become lakes of gold. The heart-shaped basal leaves are on long stalks and often have reddish undersides. The stem leaves are finely cut and ferny in appearance. The one to three foot plant is topped with a flat head of yellow daisy-like flowers about three-fourths inch in diameter. A tea made from this plant can be used as a stimulant and a diuretic.

Starry False Solomon's Seal *(Smilacina stellata)* has an unbranched terminal cluster of starry white flowers that become small, round black berries later in the season. The flowers are larger but fewer than those of False Solomon's Seal *(Smilacina racemosa)* (Plate 24), and the plant tends to be smaller and more delicate in appearance. The pointed oval leaves are alternate and clasp the stem.

Beneath the Starry False Solomon's Seal is Dog Violet *(Viola conspersa)*. This species usually holds its pale violet flowers above the foliage and may bloom when the plant is only one inch tall, although it will reach six to eight inches by the end of its growth season. The lateral petals are bearded. The base petal is veined and extends backward to form a rounded spur about one-fourth inch long. Flowers and leaves are on the same stem.

The two stalks of tiny white flowers on the right are Miterwort or Bishop's Cap *(Mitella diphylla)*. The light green basal leaves are stalked and are shaped much like small maple leaves. There is a pair of stalkless leaves about halfway up the stem. The petals of the flowers are elaborately fringed so that each flower resembles a snowflake.

At the bottom of the painting is Marsh-marigold or Cowslip *(Caltha palustris)*. This floriferous aquatic has bright golden blossoms one to one and a half inches across comprised of five to nine petal-shaped sepals. The toothed leaves are heart or kidney-shaped and are borne on long hollow stems. The flowers can be used to produce a natural yellow dye. Marsh-marigold is common in the wet areas of northern Indiana. The large, bright flowers rising above the lush, flat green leaves bloom in such numbers that they draw many people to the lakes in the spring to view their beauty.

May 1, 1985

Plate 22

The specimens for Plate 22 were found on the hills overlooking a stream near North Vernon in Jennings County. At lower left is Sweet Cicely *(Osmorhiza claytoni)*. Sweet Cicely bears sparse clusters of very small white flowers. Growing one and one-half to three feet tall, it has attractive, fern-like leaves as much as a foot or more across. The roots of this soft, hairy plant have the sweet odor of anise or licorice when bruised.

One to three small green flowers droop from the leaf axils of Green Violet *(Hybanthus concolor)*. This interesting member of the Violet family is the only eastern member of its genus. The unbranched stem grows one to three feet tall and bears alternate three to six inch leaves that are elliptic in shape and taper at both ends. While the coarse appearance fails to suggest a violet, the structure of the flower qualifies it for that family.

The Trillium *(Trillium erectum)* is commonly called Wake-robin or Stinking Benjamin. While this lusty eight to twenty-four inch Trillium bears flowers of rich color and exquisite grace and beauty, the offensive common names are well merited: it has a pronounced foul smell attractive to the carrion flies that serve as pollinators. The two-inch flowers are borne on a slender stalk above the leaves and vary in color from maroon or purple to pinkish or white. The dark green net-veined leaves are in a whorl of three at the top of the stem and may be as much as seven inches long. This Trillium can be differentiated from others of similar color by its fetid odor and its purple ovary.

Wood Anemone *(Anemone quinquefolia)* is a delicate four to eight inch woodland plant that has a whorl of three deeply cut leaves toward the top of the stem. It also has a single long-stemmed basal leaf, divided into five toothed leaflets. The pristine white, solitary flower is composed of petal-like sepals that vary from four to nine in number, although five is common.

The plant on the right side of the painting is Cleavers or Goosegrass *(Galium aparine)*. The small greenish-white flowers are borne on stalks arising from the leaf axils. The narrow, lanceolate leaves are mostly in whorls of eight and are one to three inches long. Although the plant grows eight to thirty-six inches tall, it is weak stemmed and tends to recline on the ground or on surrounding vegetation.

A land snail is shown exploring the rotten log. This individual, a particularly venturesome creature, refused to be intimidated at being handled and observed. Instead of withdrawing into its shell, it happily crawled all over us, leaving a trail of slime to mark its itinerary.

May 1, 1986

Plate 23

We discovered the specimens for Plate 23 near Turkey Run State Park. Here seepage from massive layers of bedrock provides a constant supply of moisture to the humus-rich forest litter. The steep hillsides ensure excellent drainage and protect plants from strong winds and rapid changes in temperature. A large variety of exceptionally well-grown plants are found in this friendly environment.

The pale, lavender-blue flower in the foreground is Miami Mist *(Phacelia purshii)*. This floriferous annual bears one-half inch, five-lobed flowers that are conspicuously fringed along the edges. The leaves, which are pinnately cleft, are stemless on the upper part of the plant, while leaves on the lower part of the plant have short stems. Miami Mist likes moist, well-drained areas and can grow in either shade or full sun. It tends to form colonies and often forms banks of color along country roadsides.

The Yellow Trillium *(Trillium recurvatum, var. luteum)* is the yellow form of Prairie Trillium (Plate 16). The three yellow petals are lance-shaped and stand erect. The three green sepals curve downward from the base of the stalkless flower. The stalked leaves are mottled with various shades of green and maroon. Yellow is the least common of the color forms of this Trillium.

At the top of the painting is Wild Hyacinth *(Camassia scilloides)*. The leafless six to twenty-four inch stem bears an elongated terminal cluster of six-pointed, lavender-blue flowers. The one inch flowers are composed of three slender petals and three petal-like sepals. A green bract flares out from the base of each flower. The grass-like leaves are eight to sixteen inches long and form a basal clump. The bulbs of this plant were used as food by Indians and early settlers.

On the right side of the painting is Toadshade or Red Trillium *(Trillium sessile)*. Its stalkless red flower sits directly above a whorl of three stemless leaves. The one inch flower is composed of three narrow, erect petals and three erect green sepals that spread somewhat, but do not point downward. The one and one-half to six inch leaves are mottled with light and dark green areas.

May 5, 1986

Maryrose Wampler

Plate 24

The specimens shown in Plate 24 are plants that find a favorable habitat in the alluvial soil of the floodplains produced by small streams. All were found growing along the stream at McCormick's Creek State Park.

In the foreground of the painting are the lavender flowers of Appendaged Waterleaf *(Hydrophyllum appendiculatum)*. This common biennial often covers large areas and produces an impressive spring spectacle with its dense clusters of one-half inch, bell-shaped blossoms displayed well above the foliage. Light colored stamens extend beyond the flower's corolla and give the bloom head a frothy look when viewed at close range. A hairy plant, Appendaged Waterleaf grows one to two feet tall and has lobed stem leaves about as broad as they are long. Lower leaves are pinnately divided and are more elongated.

At upper left is False Solomon's Seal *(Smilacina racemosa)*. While the plant looks much like true Solomon's Seal (Plate 18), its tiny white flowers are borne in a two to six inch pyramidal panicle at the tip of the stem rather than dangling in pairs from the leaf axils. The arched, unbranched stem grows two and one-half to three feet in length and bears alternate oval leaves three to six inches long. The leaves are hairy on their undersurface and margin and are conspicuously parallel veined. By the end of the growing season, the panicle of flowers has been replaced by a cluster of one-fourth inch ruby-red berries. This plant spreads by means of root runners as well as by seeds. It tends to form colonies and may be found in shaded woodland areas or in full sunlight along banks or in meadows.

Valerian *(Valeriana pauciflora)* is a slender, delicate plant growing fifteen to thirty inches tall. The stems bear terminal clusters of five-lobed pale pink or white flowers five-eighths to three-fourths inches long. The terminal lobe of the opposite, pinnately divided leaves is much longer than the lateral lobes.

At right is Synandra *(Synandra hispidula)*. The showy white one and one-fourth inch flowers are borne on terminal spikes in the axils of sessile bracts. The two-lipped flower has a broad, concave upper lip and a three-lobed lower lip marked inside with purple. The one to two foot plant has long-stemmed, broadly ovate leaves that are heart-shaped at the base. Neither Valerian nor Synandra are common plants, but both tend to grow in large numbers at sites where they are found.

May 10, 1986

Maryrose Wampler

Plate 25

The two plants shown in this painting, both found east of Bloomington in Monroe County, were our first experience with seeing large, colorful orchids growing in nature. Maryrose and I had read about Lady's-slippers and for years had looked hopefully for them on our spring wildflower excursions. But we had been unsuccessful, partly because orchids are not common flowers and partly because their bloom time is a little later than that of the other spring wildflowers we were observing.

Then, in a casual discussion with my students at school, I described the Lady's-slipper and our difficulty in finding it. One of my students stated that he knew where such a flower grew and would be happy to take us there. Sure enough, on a steep, wooded slope practically in his back yard were a dozen spectacular clumps of Yellow Lady's-slippers! During our excited exploration of the area, we also found a specimen of Showy Orchis.

Large Yellow Lady's-slipper *(Cypripedium calceolus, var. pubescens)* grows in rich woods and bogs. The waxy yellow pouch-shaped lip petal is one and one-half to two inches long. The lateral petals and sepals are yellow-green lightly streaked with rusty brown. Three to five pointed oval leaves may reach eight inches in length. The leaves are hairy and have pronounced parallel veining. This orchid grows nine to twenty-seven inches tall and is often found in small clumps of three or four stems.

The blossoms of the Lady's-slipper Orchids are like those of no other flower. Finding a Lady's-slipper is not a common experience, and each time I see one I wonder anew that such a flower can exist. I have heard them described as possessing lurid beauty. I'm not sure just how *lurid* relates to *beauty,* but that description rather effectively expresses the flowers' uniqueness. When I find them I feel as though I am in a magic place. They are usually scattered at random, with a spacing of ten or more feet between clumps, although occasionally a pair of plants will be found growing close together.

While none of the wild orchids can be considered easy to domesticate, Large Yellow Lady's-slipper is probably the most successful. Plants can be obtained from most nurseries that handle wildflowers. Their greatest need seems to be a constant supply of water. If the roots dry out, even when the plant is dormant, damage or loss will occur. Given a congenial setting, however, plants will persist for many years and form a large, floriferous clump in the wildflower garden.

Showy Orchis *(Orchis spectabilis)* has two broad, shining oval leaves similar to those of Pink Lady's-slipper (Plate 30) and Twayblade (Plate 18). The leafless flower stem is four to twelve inches tall and produces three to eight one inch flowers. Two lateral petals and the three sepals unite to form a mauve hood that arches over the white bottom petal or lip. A large green bract extends above each flower. These plants tend to be found in small, dense clumps.

Most Hoosiers will recognize the mushroom shown in this painting as a Morel. We were surprised to find such a choice specimen so late in the season.

May 14, 1986

Marjorie Wampler

Plate 26

All the wildflowers shown in Plate 26 can be seen at Mounds State Park. On the left is Horse Gentian or Wild Coffee *(Triosteum aurantiacum)*. Erect, stalkless red-purple blossoms are found in the upper leaf axils. The one-half to three-fourths inch flowers are cupped by five long, pointed green sepals. Egg-shaped, sessile opposite leaves are five to ten inches long. Growing two to four feet tall, this stout-stemmed plant produces red-orange berries in the fall containing three hard seeds each; the berries can be roasted, ground, and used as a coffee substitute.

At the top of the painting are blossoms of the Bittersweet vine *(Celastrus scandens)*. While many people are familiar with this plant's showy fall berries (Plate 80), few have seen its clusters of small white flowers that bloom in late spring. This twining, woody vine may grow in trees to a height of thirty feet or more. Glossy, alternate two to four inch leaves are ovate, pointed, and finely toothed. When the berries ripen, the yellow-orange jacket opens to expose a scarlet interior. Male and female flowers are usually found on different plants. I presume the flowers shown are male, since no berries developed on the vine from which they were picked. This vine was growing near enough to the ground to be easily reached. The vine with berries is growing at the top of a sassafras tree, safely out of the reach of anyone but a determined collector with a long ladder.

At upper right is Puttyroot Orchid, also called Adam and Eve *(Aplectrum hyemale)*. In summer this striking plant produces a single broad, four to seven inch leaf that is dark green with light, parallel veins. After living through the winter, the leaf dies and is usually withered or partially decomposed by the time the ten to twenty inch flower stalk emerges in late spring. The leafless flower stalk bears a raceme of eight to fifteen pale flowers about one and one-fourth inches wide. They are usually tinged purple and have a whitish, crinkly-edged lip also marked with purple. The plant rises from a soft, globular corm that is often joined to other corms by a single root. These corms lie at the surface of the soil under the forest litter and are enjoyed as food by rodents. If boiled, the corm is also palatable to humans, but because of its scarcity we recommend that it be enjoyed visually rather than gustatorially.

A Green Dragon Plant *(Arisaema dracontium)* is shown at the right. A near relative of Jack-in-the-pulpit, Green Dragon is less common and blooms later in the spring. The plant has a single compound leaf comprised of from five to fifteen pointed, dull green leaflets arranged in a distinctive semicircular pattern. A slim tipped orange spadix—presumably the dragon's tongue—extends several inches beyond the narrow spathe. Concealed at the base of the spadix are tiny, yellowish green flowers. By fall the plant has produced a tight cluster of bright red-orange berries similar to those of Jack-in-the-pulpit.

Since this painting was done in a "seventeen-year-locust" year, Maryrose included one of these interesting insects along with the case from which it emerged.

May 17, 1987

Maryrose Wampler

Plate 27

With the exception of the Blackberry brier, the flowers shown in Plate 27 have all escaped from cultivation and are not native plants. Because of their adaptability, they have been able to survive the competition from native plants and to persist without care. Dame's Rocket *(Hesperis matronalis)*, also commonly known as Sweet Rocket or Violet Rocket, resembles garden phlox in appearance, but blooms much earlier in the season and has four petals instead of five. This showy member of the Mustard family has three-fourths inch fragrant blossoms that are commonly violet but may also be pink, white, or lavender. Since Dame's Rocket is a biennial that grows from seed, a colony is frequently a mixture of colors. Broad, dark-green ovate leaves are toothed and hairy and lower leaves may reach eight inches in length. Its two to four foot height—tall for a spring-blooming plant—may be an important factor in its ability to compete in nature.

The white flowers of the Common Blackberry *(Rubus* species) are composed of five rounded petals and five pointed green sepals surrounding a tuft of stamens and pistils. Thick green canes are generously supplied with thorns that can exact a dear price for picking the juicy, delicious berries that develop for summer harvesting. Memories of blackberry picking are associated with big, iridescent green Junebugs that startle with their loud explosion into flight, and with chiggers that do not make their presence known until it is too late.

Multiflora Rose *(Rosa multiflora)* bears small clusters of three-fourths inch white, yellow-centered flowers on thorny, arching canes. The five petaled spring flowers are replaced in the fall by small, berry-like fruits called hips. As with other members of this genus, the hips are rich in vitamin C and can be made into jelly or dried and used to brew a pleasant tea. Multiflora Rose grows six to fifteen feet tall and if planted close together forms an impenetrable mass. At one time nurseries sold it for planting as a "living fence" around pastures. It was found, however, that if the pastures were not grazed or mowed for a few years, the Multiflora Rose would spread over the entire pasture. Once established, it was so difficult to eradicate that farmers soon decided they preferred dead fences to living ones. Birds and wildlife enjoy the hips and help spread the seeds. Because of its dense, thorny growth, Multiflora Rose makes excellent wildlife cover.

The type of Iris shown in the foreground has been cultivated for so long that its wild origins have been lost in antiquity. These are the old-fashioned diploid, tall bearded Irises sometimes referred to as German Iris or even given a species name, *germanica*. They are thought to be derived from natural hybrids of Eurasian origin. The flowers are composed of three downward arching, petal-like sepals called falls and three upward pointing petals called standards. A two to three and one-half foot branched bloom stalk arises from the center of a fan of erect, strap-like leaves.

The name Iris means rainbow; it reflects the wide range of colors and patterns exhibited by this flower. During the late 1800s and early 1900s many cultivars were named and introduced commercially. At first, chance seedlings were selected for their outstanding color or plant habit. Later, breeding programs were developed; specific parent plants were crossed in an effort to achieve predetermined goals. In the 1930s it was found that by using natural tetraploids—genetic accidents where the chromosome numbers were doubled—larger flowers with greater petal substance and more intense colors could be developed. Today's large, showy Irises that we see in nursery catalogs are tetraploids. Some of these gorgeous flowers would put an orchid to shame. Their only disadvantage is that they require good garden care with fairly frequent transplanting to maintain blooming vigor.

The old diploids of great-grandmother's day, on the other hand, can remain undisturbed and uncultivated for decades and continue to produce a display of fluttering flowers year after year. These are the Irises that are often seen blooming along roadsides or at old homesites years after the efforts of their original planters have been forgotten.

May 20, 1986

Maryrose Wampler

Plate 28

The flowers in this painting are often found along the edges of woods. These particular specimens were located on a rocky bank bordering a wooded lane. The yellow flower in the foreground is Trailing or Common Cinquefoil *(Potentilla simplex),* sometimes called Five Fingers because the toothed leaves have five parts. The prostrate stems of the trailing plant root at the nodes, and one plant may cover a sizable area. The five-petaled one-fourth to one-half inch flowers are at the tips of long stalks growing from the leaf axils.

The violet colored flowers are Purple Rocket *(Iodanthus pinnatifidus).* The open panicles of small four-petaled flowers vary in color from light purple to white. Leaves are stalked and the upper leaves are prominently toothed. Lower leaves may have several lobes near the base with the tip of the leaf being shaped like the upper leaves. Seed pods are long and slim and generally stand erect. Depending on light and soil conditions, this plant may be dark green or almost mahogany in color. Purple Rocket is a distant relative of the much more assertive-looking Dame's Rocket (Plate 27); both are members of the mustard family.

Fire Pink *(Silene virginica)* has bright orangeish red, long-stalked flowers one to one and one-half inches wide. The five narrow petals are often deeply cleft and the sepals unite to form a long, sticky tube. Silenes are sometimes called Catchflies because small insects may be trapped on the sticky sepals. The weak stem is six to twenty-four inches tall, but may be reclining unless it is supported by surrounding plants. Basal leaves are one and one-half to four inches long and may be lance-shaped or tapered at the base. Opposite stem leaves are sessile. Because of its bright, prominent flowers, this plant will be noticed by anyone who encounters it. It is seen often enough to make an impression, but not often enough to be familiar to most people; it is often erroneously called Indian Paintbrush because of its color.

The flower being examined by the Red Admiral butterfly is Large Yellow Oxalis or Wood Sorrel *(Oxalis grandis).* The five-petaled flower is three-fourths to one inch wide. The shamrock-like leaves often have a purple margin. The stalks that bear the seed pods are erect or spreading. This large Sorrel may grow from one to three feet tall, and except for its size and the ascending stems that bear the seedpods, it is similar to Common Wood Sorrel. The leaves, flowers, and young seedpods have a pleasant sour taste and may be added to salads or eaten by themselves.

The white head of flowers in the right foreground is Maple-leaved Viburnum or Dockmackie *(Viburnum acerifolium).* While this shrub may grow to six feet in height, it starts blooming when quite small. The three-lobed leaves, as the species name indicates, are maple-like in shape and grow two to five inches across. The flowers are prominent in the forest in late spring and early summer. In the fall, the foliage assumes a purplish pink color more attractive even than its flowers.

May 20, 1984

Plate 29

Specimens for Plate 29 can be seen at Pokagon State Park. On the left is Small Yellow Lady's-slipper *(Cypridium calceolus, var. parviflorum)*. This orchid grows eight to fifteen inches tall and bears one or two flowers whose bright yellow, waxy pouch is one inch or less in length. The spirally twisted side petals and the sepals are deep maroon streaked with green.

On the right is White Lady's-slipper *(Cypripedium candidum)*. Growing six to fifteen inches tall, this orchid has flowers with porcelain-like white pouches streaked with purple on the inside. The twisted side petals and sepals are greenish in color streaked with rust-brown.

Except for flower color, these two plants are similar enough in appearance that few people could tell them apart. Three to five alternate leaves sheath the stems at their bases. The leaves have prominent parallel veining and are pointed at the tips. Both leaves and stems are covered with fine, short hairs. These plants grow in bogs or limestone wetlands. They are often found in the light shade provided by shrubs or small trees such as Poison Sumac or Blue Dogwood.

The exotic shape of these flowers is an example of nature's variety. Made up of three petals and three sepals just like the flowers of hundreds of other plant genera, Lady's-slippers are so distinctive that they are recognized by even a novice as something special. The distinguishing feature is the lower petal, which has been enlarged into a hollow, inflated pouch. The two lateral petals are slim and spirally twisted. One of the sepals, similar in color to the lateral petals, arches above the pouch, and the other two are joined and extend downward below the pouch. These plants tend to form clumps of three or four stems, each bearing one or two flowers. A clump of Lady's-slippers is a sight of exquisite beauty.

Both of these orchids are quite rare and are on the state list of plants requiring special protection. Fortunately for visitors to Pokagon State Park, both species grow along the trail leading through the park nature preserve. Starting from the Inn parking lot, ten or fifteen minutes' easy walking along broad, well-maintained trails at the proper time of the year is all that is required to see them. These two orchids can be found on both sides of the trail, and later in the season Showy Lady's-slipper (Plate 38) can be seen here as well.

Lady's-slippers will grow only in their specialized habitat and may be dependent upon certain soil organisms that enable their roots to absorb nutrients. As a result, they must be enjoyed in their natural setting and should never be dug or picked.

Growing with the White Lady's-slipper is a Sensitive Fern, a large, coarse fern that goes dormant at the first sign of frost, leaving behind only erect, headlike fertile spikes to mark its location during the fall and winter.

May 21, 1987

Maryrose Wampler

Plate 30

The plants shown in Plate 30 were found up north, near Indiana Dunes State Park. Pink Lady's-slipper or Moccasin Flower *(Cypripedium acaule)* grows only in conditions of high acidity. We discovered these specimens in a sphagnum bog where the water was tea-colored and the growing medium was constantly saturated, but Pink Lady's-slippers will also grow in dry, sandy soil so long as the degree of acidity is adequate.

This is another orchid that must be enjoyed in its natural setting. Efforts to transplant it have been successful only in areas where a native population of plants already exists. This is thought to be due to a dependence upon certain soil fungi for nutrition. If the necessary conditions are not present, introduced plants simply linger for a year or two and then die out.

The genus name, Cypripedium, combines two words meaning "Venus" and "shoe." So while there are those who might not think of Venus as a lady, the common name Lady's-slipper is essentially a translation of the genus name. Moccasin Flower is the name usually reserved for this particular Cypripedium because the large, loose pouch looks more like a moccasin than a slipper.

Pink Lady's-slipper differs from other Cypripediums in having only two basal leaves and a leafless flower stem. The broad, glossy leaves are oblong in shape and five to eight inches long. The flower stem is six to fifteen inches tall and usually bears a single blossom. The dominant feature of the flower is the large, drooping, prominently veined pink or rose pouch, infolded or cleft down the center. Sepals and lateral petals are greenish with brown markings.

Northern Pitcher Plant *(Sarracenia purpurea)* grows exclusively in sphagnum bogs. A carnivorous plant, it supplements its nutrition by trapping insects in its pitcher-like leaves. The hollow leaves are yellow-green, heavily veined with red or purple, and have a supporting wing down one side. Standing four to twelve inches tall, they grow in a rosette from the base of the plant. The insides of the pitcher are covered with stiff hairs. Insects attracted to the lip of the pitcher find it easy to enter, but cannot exit against the downward-pointing bristles. They become exhausted and fall into the water contained in the pitcher. After they die, a combination of bacterial action and digestive enzymes secreted by the leaf produce dissolved nutrients, especially nitrogenous compounds, that are absorbed by the Pitcher Plant.

The globe-shaped flower of the Pitcher Plant blooms at the top of a leafless twelve to eighteen inch stem. Topped by three green bracts, the nodding, dark reddish purple flower is composed of five sepals and five petals that arch downward around the green, umbrella-shaped style.

Shown with the two flower species is a Tamarack bough. Tamarack, or Eastern Larch *(Larix laricina)*, is another plant uncommon in Indiana. Growing primarily in bogs and swamps, this beautiful tree needs cool summer temperatures and is occasionally found in the northern part of the state. The needle-like three-fourths to one inch leaves are pale blue-green in color and grow in circular clusters on short spurs. Although Tamarack produces cones and looks like an evergreen, it is actually a deciduous tree that turns yellow in the fall and drops its needles.

May 21, 1987

Plate 31

We located the specimens for Plate 31 in the Indian Springs area of Lawrence County. On the left side of the painting is *Iris brevicaulis* or Short-stemmed Iris. The three to five inch deep blue or blue-purple flowers are composed of three petals and three petal-like sepals. Both petals and sepals arch downward. Much broader than the sepals, the petals have a white central area veined with purple. A flash of bright yellow color flows from the heart of the flower and extends for a short distance down the center of the petals. The three style arms lie along the tops of the petals and end in short, erect, feather-like crests. The two to three-foot flower stems tend to be somewhat relaxed, becoming upright at the tips where the flowers are borne. The stalks are branched and have long, leaf-like sheathing bracts where the stems join. The arching leaves are about two and one-half feet long and up to one and one-half inches wide. Short-stemmed Iris is southern in its distribution and inhabits swamps, bottoms, and the borders of rich woods. This specimen was found in a woodland setting. We first happened across this iris a few years ago while walking through the woods on a friend's farm. We gave it a moist spot in our yard, in light shade near a downspout. It has adapted well to cultivation, increasing rapidly in size and blooming faithfully each spring.

In the center of the painting is Panicled Hawkweed *(Hieracium paniculatum)*. This native hawkweed grows from one to three feet tall and bears an open, terminal panicle of one-half inch daisy-like yellow flowers. The slender stalks are smooth and the narrow leaves are weakly toothed. As with other native species of hawkweed, leaves are found on the flower stem. In alien species (Plate 36), leaves form a basal rosette and the flower stem is leafless.

On the right side of the painting is Four-leaved Milkweed *(Asclepias quadrifolia)*. This early flowering milkweed bears one to four usually terminal umbels of pink or white blossoms. The characteristic milkweed flowers have five downward arching petals surrounding an erect five-pointed crown. The slender eight to twenty inch stem usually bears three sets of ovate to lanceolate leaves. The upper and lower sets have two opposite leaves, while the middle set consists of a whorl of four, usually larger, leaves. This milkweed differs from its sun-loving relatives in preferring a woodland habitat.

May 28, 1986

Maryrose Wampler

Plate 32

Canada Mayflower or Wild Lily of the Valley *(Maianthemum canadense)* is a small, fragrant, frothy white flower blooming in a terminal raceme above two glossy, heart-shaped leaves. This three to seven inch plant is inconspicuous singly, but creates quite a show where it forms dense carpets in the spring forest. The two leaves (occasionally three) of a blooming plant tend to be narrower and more pointed than one-leaved plants that are not yet ready to bloom. It is a true lily, although the tiny flowers are formed by two petals and two sepals rather than the usual sets of three each, giving it the appearance of being a four-petaled flower. Berries which form after the flowers are spent are light colored and are speckled. As they mature, they remain speckled but become light red.

Indian Cucumber Root *(Medeola virginiana)* is a slender-stemmed one to three foot plant with two whorls of leaves. The lower whorl, about halfway up the stem, has five to nine leaflets. The upper whorl, just below the terminal umbel of flowers, usually consists of three smaller leaflets. The sepals of the three-fourths inch flowers reflex and are pale greenish yellow in color. There may be from three to nine flowers in the terminal cluster, but only one or two open at the same time. Often they droop beneath the leaves. As the dark purple berry-like fruits develop, the stem straightens and holds them above the top whorl of leaves. The thick white edible rootstock resembles a cucumber in taste and smell. These woodland plants tend to grow in colonies. Young plants have only a single terminal whorl of leaves.

Ginseng *(Panax quinquefolium)* commonly consists of three long-stalked leaves, each with five toothed leaflets. The appearance of the leaf is much like that of a horse-chestnut; the first leaflet on each side of the petiole is smaller than the remaining three. Large old ginseng plants may be two feet tall and have five sets of leaves instead of three or four. These plants, called Five-prongers by ginseng hunters, are not common. Regardless of number, the leaves all grow from a common point at the top of the main stem. From the center of this common axil grows a short slender stalk bearing a round cluster of small, fragrant yellow-green or whitish flowers. By fall the plant will have developed a large cluster of bright red one-fourth inch berries that contain the seeds.

Ginseng roots are usually forked and have a human-like shape. Whether valid or not, many claims have been made relating the use of ginseng to longevity, vitality, and virility. Indians and pioneers used it in their herbal remedies and it has been used by the Chinese for centuries. There is still a lucrative market for dried ginseng root for both domestic use and export.

The chipmunk shown in this painting spent much of its summer lounging on the windowsill of Maryrose's studio, making occasional trips to the bird feeder for sunflower seed between naps. It ignored Maryrose as she drew and painted its portrait.

May 31, 1987

Maryrose Wampler

Plate 33

Specimens for Plate 33 were discovered along the edge of a sphagnum bog in Fulton County near Rochester, Indiana. A nearby lawn may account for the presence of the common flowers, White Clover and Creeping Wood Sorrel. The other two, Lance-leaved Violet and Large Cranberry, are not common in Indiana and are dependent on the special conditions of acidity and moisture provided by the sphagnum bog.

White Clover *(Trifolium repens)* is shown at the left side of the painting. This is the familiar clover that is frequently grown in lawns as a nitrogen-providing companion for the grass. Dense, ball-shaped flower heads may be whitish, pink, or a blending of the two colors. Leaves and flowers grow on separate four to ten inch stalks arising from a stem that creeps along the ground. While leaves commonly consist of three inverted heart-shaped leaflets, this is the famous plant that occasionally produces the "lucky four-leaf clover." The fern shown with White Clover is a young specimen of a large species called Sensitive Fern, growing to a height of two or three feet along the water's edge.

The small yellow flower is Creeping Wood Sorrel *(Oxalis stricta)*. These flowers are less than one-half inch wide and bloom singly on stalks that are scattered along the creeping stem. The seed capsule is held vertically on a stem that angles downward from its point of attachment. The entire plant may be used in salads or to produce a lemonade-like drink. Oxalis is the "sour" plant that most of us remember nibbling as children. The pleasant, mildly sour taste is produced by the oxalic acid that is found in all plant parts of any member of this genus. It is harmless in the small quantity present in the amount of sorrel needed to add flavor to a salad or brew into a glass of "sorrelade," but large amounts should be avoided.

Both White Clover and Creeping Wood Sorrel are adaptable plants. Their presence along the edge of this bog shows that they will grow almost anywhere!

In the center of the painting is Lance-leaved Violet or Water Violet *(Viola lanceolata)*, characterized by narrow two to six inch lance-shaped leaves that taper to the stalk. Blooming on separate stems, the three-eighths to three-fourths inch beardless flowers are white with magenta veining in the throat. Sometimes this veining is so concentrated, especially on the lower petal, that the flower appears to have a magenta center.

The plant on the right is Large Cranberry *(Vaccinium macrocarpon)*. Since this creeping shrub grows only about eight inches tall, "large" refers to the size of the berries it produces. The unbranched, upright stems bear numerous small oval, blunt-tipped leaves. Flower stalks arise from leaf axils along the stem and produce at their tips solitary pink flowers. The four petals recurve to form a "Turk's cap" and the stamens join to form a "beak" that gave the plant its original name, Craneberry. This is the native plant from which the commercial cranberry was developed.

June 2, 1987

Maryrose Wampler

Plate 34

The plants in Plate 34 came from sites in Fulton County. All are natives that like a dry habitat or sandy soil. At lower left is Goat's Rue or Wild Sweet Pea *(Tephrosia virginiana),* whose cheerful pink and yellow flowers, about three-fourths inch wide, are clustered at the tops of hairy stems. The pinnately compound leaves are elliptic in shape and are one-half to one and one-half inches long. Leaves usually have seventeen to twenty-nine leaflets. Seeds are in pods about two inches long and look like small peas. Abundant soft, light hairs give this plant a silvery, velvety look. Although a relative of the garden pea, this plant contains rotenone and is poisonous. Another common name, Devil's Shoestrings, describes both its poisonous nature and the plant's long, stringy roots. Indians used the crushed stems of Goat's Rue as a stunning agent to aid in the harvesting of fish.

The flower head at the top of the painting is that of Blunt-leaved Milkweed *(Asclepias amplexicaulis).* The reflexed petals are green or greenish and the upper part of the flower is usually shaded magenta. This milkweed species is easily recognized even when blooms are not present by the two to six pairs of blunt-tipped, wavy-edged leaves that deeply clasp the stem. The sturdy stalk with its large, almost rectangular leaves grows only two to three feet tall. As with other members of the genus Asclepias, young shoots and flower buds may be eaten and are enjoyed by many as a wild vegetable. The mature plant, however, is poisonous. The larvae of Monarch butterflies are able to eat fully developed leaves and retain the poison in their systems. As a result, neither the caterpillar nor the butterfly is palatable to birds, and they leave it (as well as its mimic, the tasty Viceroy butterfly) strictly alone.

Below the Milkweed is Wild Four-o'clock or Heart-leaved Umbrellawort *(Mirabilis nyctaginea).* This unusual plant has five-lobed green cups composed of joined bracts from which the flowers appear. Sometimes only one flower emerges at a time; sometimes two or three appear together. The flowers are made up of pink or purple sepals joined to look like lobed corollas. Flowers open in late afternoon and wither by the next day, thus the name Four-o'clock. The other common name, Umbrellawort, refers to the green flower cups, which greatly expand and flatten after their flowers have bloomed, suggesting an opened umbrella.

At the bottom of the painting is Hairy Puccoon *(Lithospermum croceum).* The showy golden flowers of this species are up to one inch across. The branched multi-stemmed plants may form a solid mound of gold two or three feet across. Both the stems and the lance-shaped leaves of this one to two and one-half foot plant are coarsely hairy. While its spectacular beauty would certainly qualify Hairy Puccoon as a choice garden plant, it is a short-lived perennial even in nature, and will not tolerate conditions of soil or moisture found in gardens outside its natural range.

The butterfly is a Giant Swallowtail, common in the southern states but not often seen in Indiana. It expands and contracts its range periodically, so it is more commonly found here in some years than in others.

June 3, 1987

Maryrose Wampler

Plate 35

Specimens for Plate 35 were found along the railroad tracks bordering State Highway 35 south of Winamac. In the foreground is a wand-shaped stem of Venus' Looking-glass *(Specularia perfoliata).* The one-half to three-fourths inch violet to blue flowers are found in the axils of the upper leaves. Buds in the lower axils produce seed also, but do not open into flowers. Shell-shaped, scallop-toothed leaves clasp the stem and are one-fourth to one inch wide. Venus' Looking-glass is an annual plant common in areas with poor, dry soil. It grows six to thirty inches tall.

The flowers of Smooth Phlox *(Phlox glaberrima)* are similar to those of common garden phlox. They bloom earlier in the season, however, and are smoother, more slender plants with much narrower leaves. The pink or purplish flowers are about three-fourths inch wide and have five rounded petals. Smooth Phlox reproduces itself by seed dispersal and by the formation of new plants around old crowns. It tends to form colonies, therefore, that may contain several diferent shades of color. There is also considerable variation in the width of the flower petals. Although Smooth Phlox was found at this particular site interspersed with the companion plants shown, it characteristically prefers a wetter habitat and is often found growing in ditches or low areas.

With its dense, terminal clusters of tiny white fragrant flowers, Northern Bedstraw *(Galium boreale)* is one of the showier bedstraws. Smooth twelve to thirty inch stems bear narrow, parallel-veined leaves in whorls of four. As a result of their colony forming abilities, the erect, leafy stems form sizable drifts of white among the grasses and other surrounding plants.

Wherever they grow, the blue, pea-like flowers of Wild Lupine are favorites of wildflower enthusiasts. Several different species, including the famous Texas Bluebonnet, are widespread in the western half of the United States. Only one, *Lupinus perennis,* is found in Indiana. Since Lupines are partial to dry conditions and cool nights during the summer, they are most commonly found in sandy areas of northern Indiana. One to two foot branching plants bear terminal spikes of flowers four to ten inches long. The palmate leaves have seven to nine grayish-green leaflets radiating from a center that often holds a large droplet of water after a rain or early in the morning when the dew is heavy. Hairy seedpods about two inches long mature soon after flowers fade and contain several pea-like seeds. While these seeds can be eaten if properly prepared, they contain the harmful alkaloid lupinine, and should be considered poisonous.

Lupines have deep, woody roots, and a mature plant should never be disturbed. They can easily be propagated by gathering seed that is barely ripe and planting it immediately in soil similar to that of the parent site.

The genus name Lupinus comes from the Latin word *lupus,* meaning wolf. Since Lupines tend to thrive on poor, dry soil, it was once thought that they "wolfed" or depleted the nutrients in the soil. Actually Lupines are legumes and are soil enrichers rather than soil depleters.

While these flowers possess great beauty as individual specimens, their capacity to inspire awe results from the masses of color they are capable of producing. In some parts of the country an entire landscape may be filled with them. While they are not that prolific in Indiana, they still produce banks and drifts of color in late spring that draw admirers out for a drive through the countryside to view their beauty.

The plant bearing a small cluster of tiny white flowers at the right side of the painting is Hoary Alyssum *(Berteroa incana).* The flowers have deeply notched petals. Lance-shaped leaves are one-half to one and one-half inches long. Both stem and leaves are covered with pale down that gives the plant a frosted look.

The butterfly shown in this painting is a Spicebush Swallowtail.

June 4, 1986

Maryrose Wampler

Plate 36

The flowers in Plate 36 are all plants of European origin that have found a favorable habitat in the sandy soil of north central Indiana. Shown in the background is part of the field of Christmas trees where they were found. Because of the rather poor, dry soil there, plant cover between the rows of trees is sparse. Periodic mowing, particularly while the trees are small, and soil acidity are other factors that contribute to the paucity of vegetation. These plants are well adapted to surviving in an environment that would discourage less rugged species.

At left are the bright flowers of Orange Hawkweed or Devil's Paintbrush *(Hieracium aurantiacum).* Small clusters of three-fourths to one inch flowers bloom atop leafless stems eight to twenty-four inches tall. The flower head is composed of ray flowers that have five-toothed tips. The bright yellow flower is Field Hawkweed or King Devil *(Hieracium pratense),* which grows one to three feet tall. Both of these hawkweeds have hairy, elliptic leaves growing in a basal rosette. The esteem in which these plants are held by farmers is indicated by their common names.

For some reason, now lost with the passage of time, Hawkweed was believed to be the source of a hawk's sharp eyesight. The genus name comes from the Greek root *hieros,* meaning hawk. This group of plants has been used in herbal medicine to treat eye diseases.

At top right is Night-Flowering Catchfly *(Silene noctiflora).* The deeply cleft petals of this fragrant flower are white or pink. The long, veined calyx produces a sticky substance that may trap small insects, thus the common name. Members of the genera Silene and Lychnis are similar in the appearance of both plant and flower. For those interested in such matters, the distinguishing characteristic is the three styles in flowers of genus Silene and five styles in flowers of genus Lychnis.

The small maroon flowers are those of Hound's-tongue or Beggar's-lice *(Cynoglossum officinale).* This plant is covered with silvery down and has an unpleasant mousy odor. Arched, flowering stems produce fresh blossoms at their tips while flat, four-parted fruits in various stages of development grow on the lower part of the arch. Mature seeds or nutlets are covered with hooked spines that stick to the fur or clothes of passersby.

In the foreground is Birdfoot Trefoil *(Lotus corniculatus).* The bright yellow pea-like flowers bloom in flat-topped clusters of three to six blossoms. The leaves consist of three clover-like leaflets at the tip and two smaller stipule-like leaflets at the base. The common name, Birdfoot Trefoil, is suggested by the arrangement of the slender seed pods. This legume is extremely common all over the state. It is often seen as a ground cover along roads and highways, in some places deliberately planted and in many others naturalized.

June 4, 1987

Plate 37

Specimens for Plate 37 were found in the southern tip of Indiana near the junction of the Ohio and Wabash Rivers. The purple, urn-shaped blossoms of Leather Flower *(Clematis viorna)* are shown in the left foreground. These unusual nodding flowers are made up of exceptionally thick leathery sepals that recurve at the tips. Opposite compound leaves are composed of three to seven heart-shaped leaflets. Leather Flower grows in rich soil in full sun or along the edges of woods. It is a vine that may climb ten to twelve feet high or trail over surrounding bushes and plants.

Indian Pink *(Spigelia marilandica)* is also called Pink-root, Star-bloom, and Worm-grass. These interesting common names refer either to its appearance or its use in herbal medicine for expelling intestinal worms. The one to two foot plant bears trumpet-shaped flowers in a narrow, one-sided, curving terminal cluster. The one to two inch flowers are crimson on the outside and yellow on the inside. The yellow color forms a prominent, bright five pointed star where the lobes flare outward. Rare in Indiana, this plant is the northernmost member of its genus. Most Spigelia species are found in Central or South America.

The yellow daisy-like flowers are those of Sunflower Crownbeard *(Verbesina helianthoides)*. The species name means "like a sunflower." This plant is distinguished from true sunflowers by the winged stems, and from other wingstems and crownbeards by its larger one and one-fourth to two and one-half inch flowers. The plant grows to about three feet in height and is usually unbranched. The bases of the alternate, toothed leaves run down the stem for a short distance. Sunflower Crownbeard grows along the edges of woods or in forest clearings.

American Ipecac *(Gillenia stipulata)* is shown at the right side of the painting. The five-petaled flowers are about one inch in width and may be white or pinkish. Often the narrow petals are unequal in length or are positioned irregularly around the center of the flower. The plant grows one to three feet tall and has compound leaves composed of three sharply toothed leaflets. Ovate stipules are quite large and give the leaves the appearance of having five leaflets. This is an herb that is still used medicinally as it was by both Indians and pioneers. The dried powdered root or an extract made from it is an effective laxative and emetic. Many doctors recommend that bottle of syrup of ipecac be kept in every home—particularly those with children—to be used in a poisoning emergency at the direction of a doctor or the Poison Control Center.

June 9, 1986

Plate 38

The three flower species shown in Plate 38 are Hedge Bindweed *(Convolvulus sepium),* Blue Flag *(Iris versicolor),* and Showy Lady's-slipper *(Cypripedium reginae).* The painting was begun on June 10 from specimens found in Pokagon State Park.

Considered a noxious weed, Hedge Bindweed grows almost anywhere it finds sufficient moisture. The white form of this plant is shown with Joe-Pye-Weed in Plate 70. Similar to the common Morning Glory, Hedge Bindweed can be recognized by its arrow shaped leaves with their blunt basal lobes and by its single stigma.

Blue Flag grows in marshes or wet meadows and is usually two to three feet tall. The flower is violet-blue with veining on the yellow based sepals (falls). The three petals (standards) are more or less erect and are narrower than the sepals. Blue Flag is quite variable as to intensity of color, amount of veining, and attitude of the standards. The common name is from the Middle English word *flagge,* meaning "to flutter," and originally designated rushes, reeds, and all other plants with spiky, grass-like foliage. Many people still refer to tall bearded garden iris as Flags.

Showy Lady's-slipper is an orchid with hairy leaves and stems that may grow to a husky three feet. The oval, yellow-green leaves are ribbed and sheath the stem at the base. The large flowers are starchy white, with rose-pink shading on the pouch. This orchid grows in open areas along the margins of marshes and bogs. We found these specimens growing under sparsely leaved shrubs that provided light shade. In bright morning sunlight they were difficult to see and would be easily overlooked by a casual passerby, but in shadowy evening light the rosy pouch was emphasized and the flower stood out dramatically from its surroundings.

The specimens for this painting were collected in marshy ground around the inlet of a small lake. A spring drought enabled us to walk to them with reasonably dry feet, following animal trails through the lush plant growth. While searching for our models we found that it is a mistake to position oneself between a groundhog and his den. I was engaged in selecting an aesthetically pleasing Blue Flag when rustling sounds and waving grasses announced the approach of a sizable animal. Fortunately, I happened to be straddling the trail, for a large groundhog passed right between my legs and disappeared into a hole behind me.

While collecting these specimens I experienced the first of several poison ivy-like attacks during the course of our fieldwork. I began doing research on our specimens while Maryrose started her painting. One of the bits of information I found interesting was that Showy Lady's-slipper caused a fine rash on susceptible people. Sure enough, by afternoon I had a fine rash on the inside of my forearms. By evening, Maryrose noticed that my hands were swelling and told me to remove my wedding band. It was fortunate that she did so, for by morning my hands were too swollen to grip the steering wheel of the car. Realizing that I had more than a fine rash, I checked with the park naturalist and learned that the delicate, lacy-leaved shrubs that provided light shade for the Lady's-slippers were Poison Sumac! In lifting the plug of soil containing my plant, I had coated my hands with sap from severed sumac roots.

As the doctor wrote out my prescription for cortisone pills, he told of a Vietnam War experience involving poison sumac. Men had been complaining of an ivy-like rash on their elbows and forearms but could think of no exposure to poisonous plants. The problem was traced to the local saloons. Bar tops were coated with a varnish made from poison sumac trees. Susceptible people resting their arms on the bar had an allergic reaction.

June 10, 1985

Plate 39

Pictured here are plants that might be seen on a walk beside a stream flowing through a deciduous woods. Our specimens were found at Spring Mill State Park. The little blue flower at lower left is Forget-me-not *(Myosotis scorpioides)*. Quarter-inch, sky-blue flowers with yellow eyes festoon two diverging flower branches. These uncoil as the opening blossoms progress toward their tips. This plant grows six to twenty-four inches tall and has hairy stems and leaves. Stemless, alternate leaves are one to two inches long. An escaped garden flower of European origin, Forget-me-not is a short-lived perennial that has naturalized around lakes, ponds, and streams.

At upper left is Summer Bluet or Large Houstonia *(Houstonia purpurea)*. Paired leaves are stemless, oval, and pale green in color. They have three to seven parallel veins and are three-fourths to two inches long. Pale violet flowers are long-tubed corollas that have four flaring lobes. They are borne in clusters at the ends of the four to twenty inch stems.

The one-inch flowers of Beard-tongue *(Penstemon calycosus)* are pink or purplish tubes that dilate into a wide, open-throated two-lipped corolla. The plant grows two to four feet tall and has pointed, ovate leaves. Many of the leaves are in a permanent rosette at the base of the plant. The genus name, Penstemon, refers to a specialized "fifth stamen" that also gives the plant its common name. Beard-tongues have four fertile stamens and one large, fuzzy stamen or Beard-tongue that is sterile. There are several superficially indistinguishable species of Penstemon in Indiana. All have the prominent, fuzzy stamen in the throat of the flower that is characteristic of this genus and all are known by the common name Beard-tongue. Some of these native Indiana Penstemons are large, handsome plants, but the flowers are either white or shades of pale pink or lavender and not big enough to be really showy. There are several Penstemons native to western areas of the United States that are truly spectacular, with large flowers in intense shades of blue, purple, and even scarlet. Unfortunately, they require cool summer temperatures and cannot tolerate Indiana's summer heat.

Whorls of blue-purple, two-lipped corollas rim the flowerheads of Downy Wood-mint *(Blephilia ciliata)*. A flowerhead may consist of a single whorl of flowers, or there may be three or four stacked one above the other, separated by a row of fringed bracts. Oblong, toothed leaves are almost sessile and are narrowed at the base where they join the one to three foot stem.

Single lavender-blue trumpet-shaped flowers arise from the leaf axils of Wild Petunia *(Ruellia strepens)*. These showy, five-lobed flowers are one to three inches long and have two small leaves at their base. Large, smooth leaves are paired on the one to three foot plant.

The butterfly shown on the Wood-mint is a Red Spotted Purple. The mushrooms are probably *Inocybe geophylla,* a poisonous fungus with a disagreeable odor. Color ranges from white to lilac, with differently colored specimens sometimes intermixed.

June 12, 1986

Plate 40

Some plants that like moist, acid soil are depicted in Plate 40. At the left side of the painting is Marsh-mallow *(Althaea officinalis)*. The plant was found growing in a low area bordering a woods. This narrow, columnar plant, growing two to six feet tall, is characterized by velvety, gray-green leaves that are soft to the touch. Upper leaves are coarsely toothed, and lower leaves may be heart-shaped or three-lobed. The outward facing flowers are one to one and one-half inches wide and are pink to almost white in color. This plant, introduced from Europe, was originally grown to produce a white, mucilaginous paste used to make the confection called marshmallows. While it is no longer used for that purpose, it has naturalized and continues to thrive where it finds a suitable habitat.

The tiny white flowers are those of Featherbells or Featherfleece *(Stenanthium gramineum)*. We found them at the edge of a woods bordering a tamarack bog. These three to five foot plants have a narrow, pyramidal cluster of nodding flowers one to two feet long. A member of the Lily family, the plant has six sharp pointed flower parts that are composed of three petals and three petal-like sepals. Most of the foliage is in a grass-like clump at the base of the plant. Individual leaves are eight to sixteen inches long and two-thirds of an inch wide. Flowers with female parts are found at the top, unbranched portion of the flowering stem. The branched part of the stem that makes up the tapered part of the pyramidal shape is occupied by flowers having only male parts.

At the right side of the painting is White Monarda or Basil Balm *(Monarda clinopodia)*. These plants were found on a low floodplain bordering a small, wooded stream. Growing one and one-half to three feet tall, White Monarda has smooth, ovate leaves attached in pairs to the square stem by short petioles. Most of the flower heads are borne singly at the tops of unbranched stalks. Pure white tubular flowers start blooming in the center of the flower head and progress outward. As the flower head nears the end of its bloom season, a smooth dome of bright green calyxes is left in the center of the circle of white florets.

In the foreground is Broad-leaved Waterleaf *(Hydrophyllum canadense)*. An isolated specimen of this plant looks like a twenty-inch mound of large green maple leaves. Bell-shaped white to pale purple flowers are one-half inch wide and have prominent, dark tipped stamens. The flowers bloom in clusters at the base of the upper leaf and are usually concealed by the foliage.

June 13, 1987

Maryrose Wampler

Plate 41

In this painting we see plants of European origin that are common along roads and highways in Indiana. At lower left is Crownvetch *(Coronilla varia),* given its species name because the pea-like flowers of this plant vary considerably in color, from almost white to medium shades of rose and violet. Often the upper petal is colored and the side petals are whitish. Several flowers are borne in a head-like cluster on stems that arise from the leaf axils. Pinnately compound leaves are from two to four inches long and have eleven to twenty ovate leaflets each one-half to three-fourths inch long. Numerous seeds are produced in slim, upward pointing pods. This perennial plant produces a dense carpet of creeping stems that is effective at smothering out weedy growth. It is used extensively as a groundcover on the slopes bordering major highways.

At upper right is Perennial Sweet Pea or Everlasting Pea *(Lathyrus latifolius).* The large pea-like flowers are usually magenta, sometimes pink or white. Racemes of several flowers are borne on long stems arising from the leaf axils. Gray-green leaflets are in pairs with their stems ending in tendrils. Plant and leaf stems are so broadly winged as to appear flat or ribbon-like. This escapee from the flower garden spreads by means of seeds only; the roots may produce an increasingly large specimen from year to year but remain in one place and do not send out runners.

The purple flower head on the right is Nodding or Musk Thistle *(Carduus nutans).* Nodding rose-purple flower heads from one and one-half to two and one-half inches wide are borne on long leafless stems at the end of the main stalk and its branches. Beneath the flower heads are broad, pointed purple bracts. The lowermost rows of bracts curve outward. The long, deeply lobed leaves have bases that extend up and down the stem and form wings. Both leaves and stem wings are extremely spiny. This invasive plant forms large colonies and grows from two to nine feet tall.

In the foreground of the painting is Cow or Blue Vetch *(Vicia cracca).* The downward pointing half inch flowers are borne in dense, one-sided racemes. Flowers range in color from lavender to blue and may be of startling intensity. The gray-green leaves are pinnately compound and have from eight to twelve pairs of one inch leaflets. At the end of each leaf is a pair of tendrils. Stems of this vine are about four feet long. Where no support is available to vine upon, the plant forms a mound of foliage and flowers about a foot high and several feet across.

June 14, 1987

Plate 42

Specimens for Plate 42 are plants commonly seen in meadows or along the edges of wooded areas. We found these at Brown County State Park. Blue-eyed Grass *(Sisyrinchium angustifolium)* flowers at the top of a long, flat, twisted stalk that is usually branched. A member of the Iris family, the one-half inch blue flower has three petals and three petal-like sepals, each tipped with a bristle-like point. The leaves resemble blades of grass; they are four to twenty inches long and less than one-fourth inch wide. There are several similar species of Blue-eyed Grass that are differentiated primarily on the basis of stem branching and leaf width (one is depicted in Plate 12).

The slim, stiff plants with the deep pink flowers are Deptford Pinks *(Dianthus armeria)*. Many years ago this species grew in great numbers around Deptford, England, an area that is now part of London. The one-half inch flowers bloom in flat-topped clusters at the tops of erect six to twenty-four inch stems. The flowers have five petals which are marked with tiny pin-pricks of white and have toothed margins. Slim, pointed, leaf-like bracts are found below the flowers. The narrow, light green paired leaves are one to four inches long and tend to point upward.

The Deptford Pink makes an excellent mixed border plant for the flower garden. It produces seeds in large numbers and comes up all over the flower bed. Yet its slim, airy nature allows it to grow among other perennials and bedding plants without shading or crowding them, and its height enables it to compete for light. As a result it is a charming, dainty addition to the garden that asks for no care and requires little space.

The two to three inch ball-shaped flower clusters of Common Milkweed *(Asclepias syriaca)* are borne in the axils of the upper leaves. Although the flowers of this species vary considerably in color, they are usually a shade of dusty pink. Individual flowers have five downward pointing petals surrounding an erect five-parted crown. Broad, paired oblong leaves are four to ten inches long and are light green in color. Light colored down on the undersurface of the leaves gives them a soft, thick look. Seedpods are rough-surfaced and split when ripe along one side, exposing layers of overlapping seeds. Each seed has attached a tuft of silky hairs that enables it to travel with the wind. The plant grows two to six feet tall and has a thick, unbranched stalk. Shoots, buds, young leaves, and small, firm seedpods can be cooked and enjoyed as a vegetable. The milky sap is bitter in taste and the plant contains cardiac glycosides. Both these problems are eliminated by boiling the edible parts in several changes of water. The first couple of changes should be at intervals of only a minute or two. Fresh boiling water should be used for the changes, since cold water fixes the bitter taste. Total cooking time should be about fifteen minutes.

Spiderwort *(Tradescantia virginiana)* has three-petaled, violet-blue flowers with showy yellow stamens. The three-fourths to one and one-half inch flowers are borne in terminal clusters above a pair of leaf-like bracts. Long, narrow, pointed leaves may be as much as fifteen inches in length. This plant also is frequently used as a garden flower. It forms large clumps and is very showy. Unfortunately, the flowers are open only in the morning; by noon the petals wilt and turn into a disgusting jelly-like fluid. Because of the truly outstanding beauty of the flower when fresh, many gardeners are willing to tolerate this behavior, and several named varieties are offered by the nursery trade.

June 19, 1986

Plate 43

Plate 43 shows five species of plants: Ox-eye Daisy *(Chrysanthemum leucanthemum)*, Prairie Rose *(Rosa setigera)*, Daisy Fleabane *(Erigeron annuus)*, White Yarrow *(Achillea millefolium)*, and Black Raspberry *(Rubus* species). Specimens were found in a fencerow bordering a railroad track in Monroe County.

So ubiquitous are Ox-eye Daisies that it is difficult to imagine a Hoosier summer without them. Before the coming of European colonists, however, such was the case. Native to Asia, they bloom from June through August. The flowers are about two inches across and consist of a yellow disc surrounded by fifteen to thirty white petals. The narrow, dark green leaves are deeply lobed. They form basal rosettes from which rise one to three foot stems. Ox-eye Daisies are perennials that can propagate either by dropping seed or by means of underground stems or rhizomes. The leaves are unappetizing to grazing animals, so Daisies are frequently found in pastures. Deep roots and the dense, flat rosette of leaves that shades the ground and prevents other plants from growing too close enable the Daisy to compete even against heavy grass. Most people at one time or another have played the "loves me, loves me not" game with the petals of the Ox-eye Daisy. The petals thus pulled from the yellow disc are actually complete female flowers called ray flowers. The yellow center is composed of disc flowers and contains both male and female parts that produce seed. The disc flowers open along the outer rim first and progress toward the center. Unopened disc flowers look like tiny buttons, while opened ones look fuzzy. Pollen ripens first and then the stigma matures. This sequence increases the probability of pollination from other plants, since the flower's stigma is not receptive to pollination at the same time its adjacent anther is producing pollen. The freshness of any daisy-like flower can be told by how much of the disc has bloomed.

Prairie Rose or Climbing Wild Rose is a native plant that has two and a half inch five-petaled flowers in various shades of pink. The canes may be twelve feet long and are armed with sharp, hooked thorns. Blooming in June and July, Prairie Rose frequents open woods and thickets and may be found growing up into trees. The hips or fruits of the Prairie Rose are small round, dark colored berries. The hips of any rose are extremely rich in vitamin C and may be used to brew a pleasant, healthful tea or made into jelly. The dried flower petals may also be used in herbal teas to add fragrance.

Daisy Fleabane looks like an aster with many (fifty to one hundred) tightly packed thin white or pale pink petals. The flowers are about one-half inch across and are produced in large numbers on branched stems. The lance-shaped leaves are hairy and have a toothed margin. Growing to a height of one to five feet, the plant blooms from June to October in fields and along roadsides. The name comes from an old belief that the dried flowers repel fleas. The windborne seeds germinate the same season they are produced and each plant spends the winter as a basal rosette of leaves. After it flowers the following summer, the plant dies. The ray flowers (petals) close the flower in the evening by standing straight up, perpendicular to the yellow center. The petals are a favorite food of the inchworm, the larval stage of the Geometer moth.

Yarrow is a perennial plant with soft, aromatic, fern-like foliage. The stalkless gray-green leaves are long near the base of the plant and shorten and become more sparse higher on the stem. Growing from one to three feet in height, Yarrow blooms along roadsides and in fields. This plant has a long history of medicinal use. The generic name, Achillea, refers to the legend telling of Achilles' use of this plant to heal his soldiers' wounds during the siege of Troy. The specific name, millefolium, means "a thousand leaves" and refers to the finely cut foliage. Yarrow was brought from Europe to grow in the herb gardens of the American colonies. It grows almost anywhere except in deep shade. Yarrow forms dense colonies of plants growing from rhizomes that branch repeatedly.

A member of the Rose family, Black Raspberries—also called Thimbleberries—are borne on prickly, purplish canes that are dusted with a heavy whitish bloom. Leaves are bright green on top and whitish underneath. The canes root and form new plants where their tips arch over and touch the ground. A tea made from young leaves was once used to ease labor pains and aid in childbirth. Many popular herbal teas still include raspberry leaves as an ingredient. The fruits of the raspberry are prized for jelly and pie. Many people still brave the weeds and thorny brambles along country roads and railways each summer to gather this delicious fruit.

June 21, 1984

Maryrose Wampler

Plate 44

The flowers pictured in Plate 44 were growing along a country road in Bartholomew County. On the left is Teasel *(Dipsacus sylvestris)*, a plant brought to this country in colonial times when the flower heads were used to "tease" or raise the nap of wool as part of the weaving process. Tiny lavender flowers bloom among sharp, spiny bracts on the egg-shaped flower head. Beginning in a narrow belt around the center, blooming progresses in opposite directions so that eventually there are two parallel bands of fresh flowers produced as the older flowers in the middle wither. Long, horn-like bracts curve upward from the base of the flower head, producing an artistic shape dear to the hearts of dried winter bouquet enthusiasts. Paired lance-shaped leaves are four to sixteen inches long and clasp the stem at their base. This biennial plant grows two to six feet tall and is so prickly in all its parts that gloves are needed to protect the hands when either the fresh flowers at bloom time or the dried stems in the fall are gathered.

The plant at upper right is Spotted Knapweed *(Centaurea maculosa)*. The slim, much-branched stems and thin, deeply cleft leaves give this plant an airy look. At the tops of the branches are one inch flowers in medium shades of pink or purple. Sometimes white flowers are found. Although Spotted Knapweed looks somewhat like a small thistle, there are no spines or stickers present. The spotted appearance is produced by triangular black tips on the otherwise pale bracts found on the base of the flower.

The tiny red-orange flower growing in the thin soil at the top of the outcropping rock is Scarlet Pimpernel *(Anagallis arvensis)*. These bright, star-like blossoms are borne singly on long nodding stalks arising from the axils of the paired leaves. This small annual grows four to twelve inches tall; although its flowers are usually scarlet, white or blue varieties sometimes occur. The flowers open only in bright sunlight. At night or in dull weather, they close.

The small pink, bell-shaped flowers are those of Spreading Dogbane *(Apocynum androsaemifolium)*. The interiors of the fragrant, nodding blossoms are striped with a darker pink. This shrub-like plant grows one to four feet tall and has reddish stems. The two to four inch paired leaves are oval in shape. Narrow, three to eight inch seed pods appear in pairs and split when ripe along one side to release their seeds.

At the bottom of the painting is Cheeses *(Malva neglecta);* the strange common name refers to the flat, round "cheese-shaped" seed capsules of this single species. A small creeping plant, Cheeses has one-fourth to three-fourths inch white or lavender flowers with notched petals. The scalloped one and a half to two and one-half inch leaves are round in general outline and have five to seven shallow lobes.

Shown in the background is a clump of the stately Yucca plant. Sometimes called Spanish Bayonet, *Yucca filamentosa* is an eastern species native to the Atlantic coastal plane from New Jersey south. This long-lived perennial is a popular specimen plant in gardens far to the north and west of its natural range. With its preference for poor, dry locations it is often found naturalized on rocky, well-drained soils at the tops of old road cuts. It grows and flowers best in full sun, but can tolerate the shade of shrubs and trees for many years when tall growing plants begin to encroach on its habitat.

June 23, 1987

Mariposa Wampler

Plate 45

The large canary yellow flowers shown in this painting are those of a common primrose species called Sundrops or Primrose *(Oenothera fruticosa)*. This plant is not related to the garden flowers called English Primroses. The name Primrose was probably given by colonists because the clean yellow color and the superficial appearance of the flower reminded them of the flowers they remembered from their homeland. Oenothera is a large genus and has many species that may be too similar to differentiate by casual inspection. Many open their flowers in the evening for pollination by night-flying insects. The flowers are withered by mid-morning of the next day. These species are called Evening Primroses. Several others open in the morning and remain open during the day. These species are called Sundrops. Both types of Oenothera have representatives that are popular garden flowers. Members of this genus are easily recognized by their prominent *X*-shaped stigma.

The bright yellow flowers are one to two inches wide and are composed of four broad, rounded petals that form a shallow bowl-shaped blossom. The petals are usually somewhat notched, although this characteristic varies considerably. Flowers are borne at the top of a long, slender calyx tube arising from the top of the ovary. When the flower withers, calyx tube and wilted flower drop, and the ovary develops into a strongly ribbed, cylindrical seedpod. Alternate, sessile deep green leaves are lance-shaped and have smooth edges. They are one to five inches long and have pointed tips. The perennial plants are bushy and grow one to three feet tall.

In our experience O. fruticosa is usually found growing on moist, rich soil. The specimens for this painting were found on bottomland covered occasionally by backwater from Lake Monroe. In other parts of the state, we have usually found it growing along roadsides where drainage ditches are frequently filled with water.

Lance-leaved Loosestrife *(Lysimachia lanceolata)* is shown to the right of the Sundrops. This Loosestrife can be identified by the lance-shaped, opposite leaves that taper to the base without a definite stem and by the presence of runners at the base of the plant. The five-petaled yellow flowers are about three-fourths of an inch wide and are borne facing outward or downward on long, slim stalks arising from the upper leaf axils. Petals are minutely toothed and, while rounded in general outline, have a sharp point at the tip.

This Loosestrife is often found in colonies of unbranched one to three foot plants growing along stream banks or drainage ditches. It prefers a moist location, and grows well in either full sun or light shade.

The butterfly is a Giant Swallowtail.

June 24, 1986

Plate 46

At the left in Plate 46 is Bee-balm or Oswego Tea *(Monarda didyma)*. The bright red tubular flowers of this showy Monarda form clusters from one and one-half to two inches wide. Individual flowers bloom in bands around the rim of the flower head. As the cluster matures, the center becomes a smoothly rounded dome of vacant calyxes. Dark green ovate leaves from two to six inches long are paired on square two to five foot stems. Upper leaves and the bracts beneath the flower heads may be tinged with red. Leaves are coarsely toothed and rough in texture. Bright red color and abundant nectar make this flower attractive to bees—particularly bumblebees—and hummingbirds. The leaves of this native plant were used by the Oswego Indians of New York to brew a refreshing tea.

The small dandelion-like flowers at upper left are those of Yellow Wild Lettuce *(Lactuca canadensis)*. Long, narrow clusters of one-fourth inch yellow flowers bloom along nodding stems. As the flowers mature and seeds develop, the stem of individual flowers elongates and stiffens, producing a broad, open flower and seed-bearing structure by the end of the growth season. The petite blossoms are composed of ray flowers and are open only in the morning in bright weather. This plant grows from four to ten feet tall and has oblong, deeply lobed leaves two to twelve inches long. Sharp prickles line the midrib on the undersurface of the leaves. When the hollow stems are broken, a milky sap oozes from the damaged fibers.

At upper right are the tiny pea-like flowers of Yellow Sweet Clover *(Melilotus officinalis)*. These plants, growing from two to five feet tall, bear numerous six inch tapered spikes of flowers in the leaf axils. Compound leaves are divided into three one-half to one inch ovate, toothed leaflets. The flowers are rich in nectar and attractive to bees, and smell like vanilla when crushed. The small seeds are a choice food for birds.

Intermixed with the spikes of Clover blossoms are the large, red-orange flowers of the Common Daylily *(Hemerocallis fulva)*. Leafless flowering stems three to six feet tall arise from a clump of one to three foot sword-like leaves. The stems branch near the top and have several buds, but only one or two flowers are open at a time. Although each flower lasts but a day—thus the name Daylily—the numerous buds allow a bloom season of three or four weeks. The upward-facing, funnel-shaped three to four inch flower is characteristic of the Lily family. It is composed of three orange sepals and three orange, wavy-margined petals. This plant, a native of Eurasia, was brought to this country as a garden flower. Believed by some botanists to be a natural hybrid, Common Daylily is essentially sterile and rarely produces seed. It staggers the imagination to realize that the collective acres of Common Daylilies growing here were distributed from a single, original plant. Since all parts are edible—the buds especially are prized in oriental cooking—perhaps roots or tubers were distributed by birds and animals as well as by gardeners.

Vining on the Daylily stems is Japanese Honeysuckle *(Lonicera japonica)*. The flowers are usually white at first, turning a buffy yellow as they age. Flowers bloom in pairs in the axils of the opposite ovate leaves. This voracious vine can grow thirty feet in a single season. It climbs by twining its woody stems around the trunks of trees or other upright supports. Trees supporting Honeysuckle are short-lived because of the dense, smothering evergreen foliage and because of the constriction of the trunk by the encircling vine stem. Although difficult to control, Honeysuckle has many attractive qualities. It is an effective, beautiful groundcover where its boundaries can be conveniently limited. Its fragrance has been the delight of many a twilight drive through the countryside. Many remember, as children, pulling the base from Honeysuckle blossoms to taste the droplet of nectar there.

In the foreground is Butter-and-eggs or Toadflax *(Linaria vulgaris)*. A native of Asia, Linaria was grown in European gardens as a medicinal herb and source of yellow dye. It was originally brought to this country for the same purposes, and continued to be grown as an ornamental when better sources of dye and medicine were found. This flower of old-fashioned gardens has spikes of yellow and orange blossoms. Individual flowers have two lips. The upper lip has two yellow lobes, and the lower lip has three lobes marked with orange ridges. The gray-green leaves are from one to two and a half inches long and are very narrow. Upper leaves are alternate and lower leaves are opposite or whorled. The one to three foot plant spreads by underground stems as well as by seeds and tends to form clumps. We found all of these specimens blooming along country roads in central Indiana.

June 29, 1987

Plate 47

Plate 47 was the first fieldwork Maryrose did following some knee surgery. Because she still needed to use a crutch for walks of any length, we tried to spot specimens as we drove along the roads. We were camping at Spring Mill State Park at the time, and the few flowers blooming along the shaded park roads were ones that had already been painted, so we left the park and took to the gravel roads around Mitchell. The specimens for this painting were found in a pasture that had not been recently grazed.

First we saw the Button Bush along the fence. We had found Button Bush in the northern part of the state last year and mistakenly assumed that it did not grow farther south because we were not familiar with it. As we examined the site carefully, we found the Wild Onion blooming and noticed yellow flower heads of St. John's-wort. While I was checking the St. John's-wort to see if it was a different species from the one Maryrose had already painted, I heard excited yells from the car. My attention was being directed to spiraled spikes of white flowers not twenty feet from where I was standing, which I had failed to notice. After having my attention drawn to them, I could see nothing else. I had come across Lady's Tresses only a couple of times before, and then the specimens were only eight or ten inches in height. These were over three feet tall!

Button Bush *(Cephalanthus occidentalis)* is an aquatic shrub with small flowers that form a ball about one and a half inches across. The ovate leaves are three to six inches long and have an opposite or whorled position on the stem. Growing from three to ten feet tall, the plants are found in swamps and along the banks of ponds and streams. This specimen was growing in a shallow ditch between the road and a fencerow. The common name comes from the ball-like flowers and fruit heads. The fruit heads provide a source of food for wildlife; they are especially enjoyed by Mallard ducks.

Spring-flowered Spiral Orchid *(Spiranthes vernalis)* is one of the numerous species of Lady's Tresses. There are four to five grass-like leaves on the basal part of the stem. As many as fifty flowers, each about half an inch long, line the stem in a spiral or along only one side. Flowers are whitish with a yellow center on the lip. This plant was growing in full sun in a damp meadow. We found it interesting that one of the flower spikes spiraled in the opposite direction from the others.

The Alliums were found growing in both pasture and fencerow. The Wild Onion *(Allium stellatum)* has a showy umbel of six-pointed, lavender flowers. Field Garlic *(Allium vineale)* has the flowers mixed with bulblets, or flowers may be totally absent, being replaced by bulblets with "tails."

Heal-all or Selfheal *(Prunella vulgaris)* is a member of the Mint family that was once used as a cure for throat ailments. This perennial plant has escaped from herb gardens and has naturalized almost everywhere. The light to dark purple flowers are about one half inch long. The upper lip is arched, the lower lip is drooping and fringed. It is also lighter in color. The leaves are opposite and vary from lanceolate to ovate and from smooth to toothed. The plant grows from six to twelve inches tall and sometimes sprawls. Blooming from May to September, it is commonly found in gardens, fields, and roadsides.

Fogfruit *(Lippia lanceolata)* is a member of the Verbena family. It grows along stream banks and in wet forests and meadows. When it first starts blooming, Fogfruit flower heads are almost round. As flowers open upward toward the tip, the spike elongates to as much as fifteen inches. Sometimes Fogfruit is listed under the older genus name *Phyla*.

July 2, 1986

Maryrose Wampler

Plate 48

The native plants shown in Plate 48 can be seen in the Indiana Dunes State Park area. At left is Prairie False Indigo or White False Indigo *(Baptisia leucantha)*. This bushy, shrub-like legume grows two to five feet tall and has grey-green leaves palmately divided into three smooth, oval leaflets. Open, erect racemes of one inch pea-like white flowers may reach as much as twenty-four inches above the foliage. The large spikes of white flowers stand out dramatically against a dark background of open woods or roadside banks where these plants frequently grow.

The common name False Indigo results from the blue dye contained in this and several other species of Baptisia. While there is a blue undertone to the stems and occasional shadings of blue on the flowers, the blue dye content is not apparent until the plant is dried, whereupon all plant parts appear black as a result of the dye concentration.

In working with this plant, we found that it does not like to be disturbed. Although the beautiful spikes of flowers would lend themselves well to bouquets, we found that the flower promptly wilts unless cut at dusk or very early in the morning. Massive, woody taproots make attempts at digging the plant a destructive waste of time. Fortunately for those interested in wildflower gardening, the pea-like seeds are easily gathered and can be planted where this distinctive plant is to be positioned.

In the center of the painting is Rattlesnake-Master or Button Snakeroot *(Eryngium yuccifolium)*. Smooth stems two to six feet tall branch near the top and bear terminal, thistle-like flower heads. The rounded three-fourths inch flower heads bear small, unobtrusive greenish-white flowers mingled with stiff, pointed bracts. Lower leaves, which may be as much as three feet long, are narrow and have bristly edges. The leaves are similar in character to those of a Yucca and are the basis for the species name yuccifolium.

The strongly reflexed petals of Michigan Lily *(Lilium michiganense)* suggest the common name, Turk's-cap. The Turk's-cap Lily Maryrose had originally intended to paint was *Lilium superbum*. The two species are quite similar except for the larger flower and prominent green, star-shaped marking in the throat of L. superbum. In our travels around the state, we found Michigan Lily to be fairly common along railroads and roadsides in Northern Indiana. The only natural site we have ever found for L. superbum was along a railroad near our home in Monroe County. This site was eliminated when weeds along the track were sprayed with herbicide.

Michigan Lily commonly grows three to six feet tall and has whorled, smooth-edged, sword-shaped leaves. Leaves on the lower part of the stem may be alternate. On a strong plant, the flowering stem branches and may support as many as twenty two to two and one-half inch blossoms. The recurved petals and sepals are usually red-orange at the tips shading to gold or yellow at the throat. The central area of the flower bears numerous cinnamon-colored dots.

The plant with the huge oval leaves shown growing near the railroad track in the background, just behind the False Indigo, is Prairie Dock (Plate 54). Because of the size of the plant, it was not possible to show the basal leaves with the flowers in that plate.

July 3, 1987

Maryrose Wampler

Plate 49

Plate 49 depicts plants found at Tippecanoe River State Park, growing near a marsh. At one time, much of the land in this part of the state was too wet for farming. Systems of ditches were dug to drain the marshes and to provide for sufficient drying of meadows to allow for the planting of crops. In some areas of the park dams have since been constructed to restore the original marsh conditions. Such an area is shown in the background of this painting.

At the left is Flowering Spurge *(Euphorbia corollata),* whose small, pristine flowers are composed of five rounded, white bracts surrounding minute flower clusters. The long, smooth oval leaves are bright green in color and are alternate except for a whorl of several leaves at the base of the flower head. The one to three foot stem exudes a milky fluid when broken. Flowering Spurge is found in fields, open woods, and along roadsides throughout most of the state. A well-grown plant may form a mound of white two or three feet in diameter that resembles the garden flower Baby's Breath. This plant is a powerful cathartic and has been used in herbal medicine as a laxative. With improper dosage, it would be considered poisonous.

Next to the Flowering Spurge is Pale Spiked Lobelia *(Lobelia spicata).* The small pale blue to white flowers have the characteristic Lobelia shape. The corolla has three prominent lobes extending downward and two smaller, less conspicuous lobes that point upward. The plant consists of an unbranched stem with flowers clothing the upper third and nearly toothless sessile alternate leaves occupying the lower portion.

Colicroot or Star Grass *(Aletris farinosa)* is a spike of whitish flowers that resemble Lady's Tresses (Plate 47) in appearance. The urn-shaped flowers are one-fourth to one-half inch long with a granular texture. This plant is a member of the Lily family and has the characteristic six-parted flower. Most of the long, narrow pointed leaves are in a rosette at the base of the plant. The few stem leaves are small and bract-like. Colicroot grows in meadows and open woods and likes acid soil. Until the nineteenth century it was used in herbal medicine to treat colic.

The stem of yellow flowers is Hairy Hawkweed *(Hieracium gronovii).* This native plant is characterized by a long panicle of flowers, its length three to ten times its width. The two to six inch alternate leaves are hairy in texture and oblong in shape. Leaves are found on the lower half of the one to five foot stalk. The presence of stem leaves differentiates native Hawkweed species from common alien species, which have a bare flower stem arising from a basal rosette of leaves.

The light purple flowers at upper right are Canadian Thistle *(Cirsium arvense).* The numerous ball-shaped, fragrant flowers are one- half to three-fourths inch wide. The spiny leaves are five to eight inches long with deeply cut, wavy edges. Growing one to five feet tall, this plant spreads rapidly and forms large colonies that are difficult to eradicate. Our most common thistle, Canadian Thistle is a native of Europe that came to the United States by way of Canada, hence the common name.

Although this thistle rates high on the "most hated plant" list, it is a beautiful flower early in its bloom season before seed heads start to form. Its heavy, sweet fragrance draws hordes of insects and butterflies. As the fluffy heads of seed develop, goldfinches search it out. They delay their breeding season until thistledown is available to use as a lining for their nests. Thistle seed is also one of their favorite foods.

The frothy lavender flower in the foreground is Narrow Leaved Mountain Mint *(Pycnanthemum tenuifolium).* The one to two inch leaves are very narrow. The numerous tiny flowers bloom in branching, flat-topped clusters on one to three foot plants.

July 5, 1985

Plate 50

Specimens for Plate 50 were found at Martin State Forest in southern Indiana. Five widely spaced petals alternating with five shorter green sepals comprise the flowers of White Avens *(Geum canadense).* As the flower matures, it becomes a bristly, round seed head about five-eighths inch in diameter. The large, attractive lower leaves are usually divided into three palmately compound sections. The one to two and one-half foot stem bears simple leaves on its upper section.

The numerous racemes of blue flowers are those of Downy Skullcap *(Scutellaria incana).* The three-fourths to one inch flowers have an arching, hooded, upper lip and a flaring lower lip with a white area in the center. On the top of each calyx is a tiny skullcap-shaped growth that gives the plant its common name. The upper part of the plant is hoary with short whitish hairs. The ovate leaves are long-stalked, coarsely toothed, and pointed at the tips.

Blue flowers are especially welcome in early summer when yellow ones dominate the landscape. Skullcap's elegant blooms, set against attractive, bright green foliage, are particularly beautiful in the dappled shadows of forest pathways. While Skullcap may be found situated in bright light, it is one of the few summer flowering plants that can grow and bloom in deep shade.

At the bottom of the painting are the fully expanded leaves of Goldenseal or Yellowroot *(Hydrastis canadensis).* The plant is normally composed of three large leaves, one basal leaf and two leaves on the flowering stem. These five to seven lobed leaves reach six to eight inches in width and are prominently veined. Each blooming plant bears a single large red berry on a stem so short that it appears to be nestled on the surface of the top stem leaf. A colony of these plants looks like Christmas in summer, with the bright red of the raspberry-like fruits ornamenting the rich, dark green of the leaves. The flower of this spring blooming plant is shown in Plate 20.

In the course of reading about the plants Maryrose has painted, we have encountered many medicinal plants that were considered cure-alls. Ginseng and Goldenseal are about the only ones that continue to have monetary value. We wonder if most of these plants' curative powers haven't been psychological. Many make a pleasant or at least an interesting tea, and some of them taste so awful that people probably thought they must be good for whatever ailed them. While Maryrose was doing the preliminary work for this painting we tried brewing a tea from Goldenseal roots. The tea had a beautiful light chartreuse color and a strong, unpleasant "green" taste. Most of the curative claims made for Goldenseal are related to the treatment of stomach ailments. It was indeed soothing to our stomachs, but it irritated the lining of our mouths. Since many medicinal herbs are toxic if consumed in excessive amounts, we decided not to experiment further unless we could find more information about dosage.

The ladder-like leaves of an Ebony Spleenwort fern are shown above the Goldenseal. This woodland fern grows about one foot high and tends to be found singly or in small groups. It is easily recognized by its shape and by its distinctive shiny dark stems.

The butterfly shown on the Skullcap is a Pipevine Swallowtail.

July 5, 1986

Maryrose Wampler

Plate 51

Through most of the summer the large, colorful wildflowers of the genus Hibiscus ornament lake shores, swamps, and drainage ditches in many parts of Indiana. Plate 51 shows Crimson-eyed Rose-mallow *(Hibiscus palustris, forma peckii)* blooming along the shore of a small lake. This coarse three to eight foot plant bears five-lobed four to seven inch flowers that are white with a red or maroon "eye" at the center. The long, curving style is tipped with five velvety stigmas, and has numerous golden stamens radiating from its lower third. Narrow leaf-like bracts are present beneath the large, green calyx. The ovate, yellow-green leaves are four to six inches long and have toothed margins and pointed tips. The leaves are downy on their undersurface.

In the foreground is Lizard's Tail *(Saururus cernuus),* an interesting flower that has neither sepals nor petals. The six inch flower spike with its drooping tip bears numerous small white flowers composed of three to four pistils united to form a central column surrounded by six to eight showy stamens. The dark green, heart-shaped leaves are three to six inches long. This two to five foot plant grows in swamps or shallow water.

In the background is White Water Lily *(Nymphaea odorata).* This aquatic plant has white or pink fragrant flowers that are three to six inches across. The many petals are interspersed with numerous yellow stamens. Inner rows of petals become progressively narrower. Flowers open in the early morning and close at mid-day. The flat, floating leaves are four to twelve inches in diameter and are deeply notched at the base; they look like a pie with a narrow wedge missing. The leaves are a shiny green on top and purplish-red on their undersurface. As seeds begin to grow the flower stem coils, drawing the seedpod beneath the surface of the water where the seeds mature and ripen. When they are released from the pod, they float for a time before sinking to the bottom to germinate. These seeds were used by Indians and pioneers for food and are eaten by waterfowl and other wildlife.

We found these specimens in the vicinity of Chain O'Lakes State Park. The lakes there, called kettle lakes, were formed at the close of the last glacial period by the melting of ice blocks buried in glacial till. Since the flowers shown here actually grow in the water it is often difficult to get close enough to them along the lake shore to enjoy their beauty. If bird watching is being combined with a wildflower trek, binoculars can be used to get a good close-up look. The most effective and enjoyable way to examine these flowers closely, however, is to rent one of the state park rowboats and approach them from the water instead of the land.

July 6, 1987

Plate 52

The plants in Plate 52 are ones that might be seen while walking through open woods or along wooded roadsides. These were found at Tippecanoe River State Park. At lower left is Horse-mint *(Monarda punctata),* whose wide-jawed, yellowish, purple-spotted flowers are in dense whorls surrounded by whitish or lilac colored bracts. Individual flowers are three-fourths to one inch long. The bracts are so conspicuous that, when viewed from any distance, Horsemint gives the impression of being lilac and white. The lanceolate leaves are opposite and shallowly toothed. This plant grows one to three feet high and prefers dry, well-drained soil.

At upper left is Meadowsweet *(Spiraea latifolia).* This woody shrub grows two to five feet tall with dense, pyramidal clusters of small white or pinkish flowers at the tips of its branches. The reddish or brownish stems bear oval to broadly lanceolate leaves that are one and a half to two and three-fourths inches long. Leaves are coarsely toothed, and both leaves and stems are smooth in texture. After flowering, conspicuous brown seed heads develop that persist into winter. Meadowsweet is tolerant of wet soil conditions and may be seen growing among cattails as well as in drier habitats.

The center of the painting shows the spiked bloom head of Leadwort or Prairie Shoestring *(Amorpha canescens).* The tiny, dark purple flowers have prominent bright orange stamens. A dense covering of short white hairs gives this plant a light gray-green color. The two to six inch leaves are pinnately compound and have fifteen to forty-five small leaflets. Growing one to three feet in height, Leadwort has long, slim roots that reach four feet into the soil—the inspiration for its common name, Prairie Shoestring. These roots protect the plant against drought and enable it to compete with shallower rooted plants and grasses for moisture. The Indians used the leaves of Leadwort as a tobacco substitute and for making tea.

The loose panicle of pink flowers above the Leadwort is Pointed-leaved Tick Trefoil *(Desmodium glutinosum).* Growing one to four feet tall, this plant has a whorl of leaves at the base of the flower stalk. Leaflets are broad and pointed at the tip. The several species of Tick Trefoil all produce numerous flat, fuzzy seed cases that break into triangularly shaped segments that stick to the clothing (or fur) of passersby. New colonies of Tick Trefoil are planted where these travelers are picked off.

At the right is Germander or Wood-sage *(Teucrium canadense),* whose pink three-fourths inch flowers are borne in a terminal spike and consist of a five-lobed corolla. The bottom lobe is long and broad while the upper and lateral lobes are much shorter and smaller. The two to four inch leaves are opposite and may have small secondary leaves at the axils. They are toothed, lance-shaped, and have short, dense hairs on their undersurface. Germander grows one to three feet high. It is another of those plants that we at first thought grew only in northern areas but have since found almost everywhere in Indiana.

July 7, 1985

Plate 53

Specimens for Plate 53 came from a moist area along one of the roads in Tippecanoe River State Park. At the left is Monkey-flower *(Mimulus ringens).* The genus name as well as the common name reflects the amusing, face-like shape of this flower, *mimulus* being the Latin word for a buffoon. The light purple flower is about one inch long and consists of two lips. The upper lip has two lobes and the larger, bottom lip has three lobes with two yellow spots on the inside. The slender flower stalks are longer than the tube-like calyx and arise from the upper leaf axils. Toothed, sessile leaves are two to four inches long and are oblong to lanceolate in shape.

Above the Monkey-flower is Steeplebush or Hardhack *(Spiraea tomentosa).* This small, erect shrub has dense, steeple-shaped panicles of tiny pink five-petaled flowers that begin blooming at the top of the panicle and progress downward. The toothed, oblong leaves are one to two inches long. Their woolly undersurface is a brownish or rusty color. Sometimes seedpods persist from the previous year and can be found on the plant along with fresh bloom heads.

The flowers of Blue Vervain *(Verbena hastata)* first bloom at the bottom of the stiff, candelabra-like spikes and progress upward. The small blue-violet five-petaled flowers bloom a few at a time in a circle around each spike. The four to six inch opposite lanceolate leaves are doubly toothed and rough in texture. Because Verbena was thought to be a cure-all in ancient times, the genus name is Latin for "sacred plant." Bumblebees are important pollinators for this sturdy perennial.

Below the Vervain is Starry Campion *(Silene stellata).* The three-fourths inch flower has five fringed white petals. The green sepals unite to form a five-lobed bell. The smooth, lance-shaped leaves are often in whorls of four.

The heads of lilac flowers on the right are Monarda or Wild Bergamot *(Monarda fistulosa),* a perennial plant that spreads eagerly both by underground stems and by seeds. It can form large areas of color ranging from pinkish to lavender. Bracts beneath the two inch flower heads are often tinged with pink or purple. The coarsely toothed leaves are triangular or lance-shaped and are rough in texture. The common name, Bergamot, was given this Monarda because its citrusy-minty fragrance reminded early settlers of the smell of bergamo oranges grown near Bergamo, Italy. As with most Monardas, its leaves can be used to brew a refreshing tea.

July 9, 1985

Plate 54

The plants shown in Plate 54 are characteristic of prairie areas in Indiana—places that tend to have fairly dry, well-drained soils, often sandy or gravelly. Sometimes prairie conditions exist where only a thin layer of soil covers the bedrock. These conditions are found primarily along the western side of the state. Another excellent place to look for prairie flowers is along old railroad tracks, whose raised bed and gravel fill produce a favorable habitat for many interesting and beautiful plants. Often such sites exhibit a much more varied community of plants than the surrounding countryside. These particular specimens were found along the track that runs beside State Road 35 south of Winamac. This stretch of road for fifteen or twenty miles is one of the most spectacular "flower beds" I have ever seen. It is beautiful through the seasons, but is most showy in early summer when the predominantly yellow flowers are in full swing.

The tall flower on the left is Prairie Dock *(Silphium terebinthinaceum)*. Only the top portion of this huge plant is shown. Most of the oval or heart-shaped leaves are at the base of the stalk and may be six to ten inches wide and as much as two feet long. Basal leaves of this plant are shown in the background of Plate 48. The flowering stalk arising from the basal rosette of leaves is four to ten feet tall. The top of the flowering stem branches and supports several two to three inch yellow daisy-like flowers. The buds also are of interest, looking something like large marbles at the tops of the slender branch tips. This plant is well adapted to dry soils; it has a taproot that reaches down several feet. A seedling two inches tall will have a thick taproot a foot long. Obviously, anyone wanting to grow this spectacular plant needs to collect seeds rather than trying to transplant it!

In the center of the painting is Gray or Prairie Coneflower *(Ratibida pinnata)*. A rough, hairy plant, Prairie Coneflower grows three to five feet tall and has pinnately divided leaves about five inches long. The daisy-like flowers are composed of drooping, bright yellow ray flowers (petals) and a grayish central cone that becomes dark brown as the disc flowers open. The disc flowers bloom in a single line around the cone, beginning at the bottom and progressing toward the top. The age of a flower can be determined by how much of the cone has bloomed. You might think the drooping petals would make this plant seem wilted or discouraged; on the contrary, they give it a perky appearance. Blooming in masses of sunny color, it looks like a corps of ballerinas dancing across the landscape.

At the right is the third yellow daisy-like flower in this painting—Rosinweed *(Silphium integrifolium)*. Growing two to five feet tall, this plant has sessile leaves that may have either opposite or alternate positions on the stem. Surrounded by leafy bracts, the buds open into flowers one to two inches wide. Rosinweed is so named because of the gum-like resin that oozes from the stems when a flower head is broken off. This resin when dried may be used like chewing gum.

At the bottom of the painting is Pasture Rose *(Rosa carolina)*. This rose has the usual five-petaled "wild rose" flower. The slender stems have straight thorns and grow upright to a height of one to three feet. Since the plant spreads by means of underground stolons, it tends to form colonies. Pasture Rose blooms for several months beginning in late spring and so may have both blooms and hips (rose fruits) at the same time. As with any rose, both petals and hips may be used for food.

July 15, 1985

Plate 55

The pendulous yellow bell-shaped flowers at lower left in this painting are those of the Ground Cherry (*Physalis heterophylla*). Flower stalks bearing their solitary, five-lobed blossoms grow from the leaf axils. The alternate leaves are heart-shaped and have wavy margins. The much branched plant grows one to three feet tall, but being weak stemmed may sprawl on the ground with only the branch tips upright. This deep-rooted perennial bears interesting tomato-like fruits enveloped in a parchment-like calyx. While the plant and green fruit are poisonous, the ripe yellow fruit is commonly made into jams and pies and serves as a food for insects, birds, and rodents in the fall. The popular garden perennial Chinese Lanterns is an alien member of the same genus. They look very much alike, but Chinese Lanterns has a more upright habit of growth and a larger, bright orange calyx.

At upper left is Sow Thistle *(Sonchus asper)*. Its flowers look like small dandelions, one-half to one and three-fourths inches wide, and mature to a fluffy seed head also similar to that of a dandelion, but smaller. The spiny leaves are curled and convoluted, and clasp the stem at their base. This one to six foot plant is an annual, coming up from seed each year and dying after one year's growth.

Moth Mullein *(Verbascum blattaria)* may be either yellow or white, with orange anthers and purple shadings. The one inch flowers and button-like buds are prominently displayed in spike-like clusters at the tops of single, erect stems. The coarsely toothed leaves are triangular in shape and clasp the stem at their base. The common name, Moth Mullein, is suggested by the similarity of the orange anthers to a moth's antennae.

Goatsbeard or Meadow Salsify *(Tragopogon pratensis)* with its yellow flower and huge seed head is at the right side of the painting. The yellow rays of the one to two and a half inch flower are supported by long, pointed green bracts. Grass-like leaves clasp the stem. This plant has an edible root and is related to the cultivated Salsify or Oyster Plant. The beautiful seed heads are sometimes used in dried flower arrangements. They can be preserved intact if given a gentle mist of hair spray. Another method of preserving them is to reach into the center of the seed head with a wire or slender stick and anchor the seeds with droplets of model airplane glue.

The yellow flowers with five notched petals at the bottom of the painting are Rough-fruited Cinquefoil *(Potentilla recta)*. Beneath the five petals are five green sepals, and below those, usually smaller and alternating with them, are five green bracts. This one to two foot hairy plant is very leafy and much-branched. The coarsely toothed leaves are divided into five to seven rather narrow leaflets. As a result of the shape of the leaves the plant is sometimes called Five Fingers.

The unobtrusive insect resembling a twig that is shown on the Ground Cherry is a Walkingstick.

July 15, 1985

Plate 56

The specimens for Plate 56 were found along the entryway to Harmonie State Park. Maryland Golden Aster *(Chrysopsis mariana)* bears yellow aster-like flowers, but is not a true aster. The flowers are about an inch wide and bloom at the tips of the much-branched stems. Sticky bracts are located below the flowers. The oblong, sessile leaves are one to two inches long and are toothless or have only a few shallow, widely spaced teeth. The plant is rough in texture and both stem and leaves are covered with short, stiff hairs. Maryland Golden Aster prefers sunny, dry, well drained growing conditions and is usually found in sand or along the gravelly edges of roads. Although not considered rare, it is not a common plant in our experience. We had never seen it growing in a natural site except in this one location. Then, quite recently, I identified a plant growing along the highway a few miles from home. Since I had never seen it there before, I wonder if it could have been a seedling from my own wildflower garden.

Wild Potato-vine or Purple-throated Morning Glory *(Ipomoea pandurata)* is a trailing or climbing plant that may reach fifteen feet in length. The gorgeous two to four inch flowers are white with a prominent purple or magenta area in the throat. They are fresh and crisp in the morning, but wither by noon each day. Heart-shaped leaves are two to six inches long. As lovely as it is, this plant is generally considered a noxious weed because its rampant growth and abundant, broad leaves cause it to smother the plants it vines upon. Wild Potato-vine grows from a perennial root that may weigh as much as fifteen pounds. While edible, it has a bitter taste that discourages its use as a food source except in times of dire need.

Both of these plants grew on a sand hill overlooking the Wabash River a few hundred yards from the house where Maryrose grew up, and for her they are inextricably linked with the sandburrs which also grew there. One of her strongest childhood memories is of playing barefoot in that area and stepping on one of the abundant sandburrs. She says the only thing worse was landing on another burr when she sat down to remove the one she had stepped on. Maryrose therefore felt this painting would be incomplete without a sandburr, but we neglected to collect a specimen when the painting was started. In the fall when we returned, we could not see any. After carefully searching areas where we had found plants growing earlier in the season, I finally located a cluster by running my hand over the thatch. First removing from my fingers three burrs with spines that were barbed like fishhooks, we collected the plant and brought it back to the studio. Everything Maryrose had said about sandburrs was true. The hard, woody spines are so strong and sharp that they can penetrate a rubber-soled shoe!

July 18, 1986

Plate 57

The plants shown in Plate 57 can be found at Ouabache State Park. On the left is Trumpet Vine *(Campsis radicans)*. This showy native vine with its bright red-orange trumpet-shaped flowers is prominent along roadsides and fencerows in much of Indiana. Long lived and woody-stemmed, it develops a trunk capable of supporting itself after several years' growth. Sometimes a beautiful specimen plant is produced where a vine has climbed a fencepost and branched out laterally at the top. With the passage of years the post rots away, leaving a short, stout vine trunk with an umbrella of flowering branches at the top. This plant is so ornamental that it is commonly offered in nursery catalogs as a cultivated vine. But one of its common names, Hell Vine, effectively expresses local attitudes in places where much time is spent trying to hoe it out of the back yard vegetable garden. The vibrant flowers of Trumpet Vine consist of five-lobed corollas about three inches in length. One or two flowers open at a time from clusters of buds at the branch tips. Pinnately compound leaves are eight to twelve inches long and have seven to eleven toothed leaflets.

St. John's-wort *(Hypericum perforatum)* is a plant of European origin that is widely distributed. Growing one to two and a half feet tall, numerous slender branches support sunny yellow flowers about an inch across. The opposite, sessile leaves are oval in shape and are one and a half inches or less in length. The species name, *perforatum,* refers to translucent dots that are apparent when leaves are held to the light. The five-petaled flowers have tiny black dots along the petal margins. St. John's-wort is a common plant along roadsides and in fields that have not been plowed for several years.

Also of European origin, Common Mullein *(Verbascum thapsus)* is found in virtually all parts of the United States. This regal plant rises two to eight feet from a rosette of oblong four to twelve inch flannel-textured leaves. The semi-woody stem is unbranched, although the flower head may divide into several candelabra-like spikes. The yellow five-petaled flowers are three-fourths to one inch wide and are irregularly scattered among small leaves, buds, and developing seed capsules on the closely crowded flower stalks. A biennial, Common Mullein develops an attractive cabbage-like cluster of thick, fuzzy, grayish leaves in its first year of growth. It blooms the second year; then, after producing seed, it dies, although the woody nature of the stem may cause it to persist as a blackened stalk into the third season.

Thimbleweed *(Anemone virginiana)* is a common native anemone growing two to three feet tall. The slender stem arises from compound basal leaves and has a whorl of two or three smaller compound leaves about half way up the stalk. At the top of the stem is a single white flower one to one and a half inches wide. The "petals" are actually white or light green sepals. The number of petals may vary from four to nine although five is usual. The common name is suggested by the thimble-shaped seed head that develops as the plant matures.

July 20, 1985

Plate 58

The plants depicted in Plate 58 can be seen in the area of Lincoln State Park in southern Indiana. The plant in the left foreground is Hoary Vervain *(Verbena stricta)*. This plant is shorter, more generously foliaged, and has larger flowers than its rangier relative, Blue Vervain (Plate 53). The flowers are more violet than blue, sometimes approaching a true pink. This attractive wildflower is similar in appearance to Veronica, the cultivated garden flower. The thick, coarsely toothed ovate leaves, almost stalkless, are frosted with a heavy covering of whitish hairs. The gray-green color of the one to four foot plant provides an effective setting for the showy, candelabra-like spikes of flowers. Bumblebees are attracted to this flower and are important pollinators. A European relative of this native American wildflower was considered an herbal cure-all in ancient times. As a result the genus name means "sacred plant."

The stem of ripe blackberries belongs to the genus *Rubus*. This genus contains multitudinous species that are difficult to differentiate. As a youngster I remember picking blackberries for two dollars a gallon—big money for a preadolescent in those days. I enjoyed roaming sunny, brier-infested meadows and thickets along the edges of woodlands and was an enthusiastic berry picker. On a few occasions, though, snakes and I surprised each other in the underbrush. After seeing one, I was sure that any sound I heard or any movement I saw was a snake. While I was interested in snakes and played with harmless ones on occasion, the imaginary snakes that infested the brier patch after I had seen a real one—harmless or not—were all copperheads and rattlesnakes. On these occasions, I went home early.

Queen Anne's Lace (*Daucus carota)* is also called Wild Carrot, because that is what it is, or Bird's Nest, because of the cup shape the flower head assumes as it dries. The two to five inch flower clusters are extremely flat and form a lace-like pattern of tiny, creamy-white five-petaled blossoms. Five anthers protrude from each flower and add to the frothy appearance. There is often a deep purple flower in the center of the cluster. Stiff three-forked bracts are found at the base of the flower heads. The two to eight inch leaves are very finely cut, giving a fern-like appearance. The stem is two to four feet tall and is covered with bristly hairs.

Queen Anne's Lace is a biennial plant that forms a rosette of foliage the first year. At this stage the roots can be eaten like the cultivated carrot that was developed from it. If the plant achieves sufficient size during its first year of growth, it sends up one to several stalks and blooms the second year. After blooming the plant dies, although the stems often remain standing and disperse seeds throughout the winter. This plant is often a host for the larvae of Black Swallowtail butterflies.

Chicory *(Cichorium intybus)* is sometimes mistakenly called Cornflower because of its color. Although Maryrose had already painted Chicory in Plate 65, she decided to repeat it in this picture because in subsequent seasons we found it so often—and so beautifully—growing with Queen Anne's Lace.

The butterfly shown in this painting is a female Spicebush Swallowtail. Females have an area of blue on the hind wing, whereas males have green.

July 20, 1986

Plate 59

The specimens for Plate 59—all prairie flowers—were found at a place where the soil has never been plowed—a small cemetery dating from pioneer days. Royal Catchfly *(Silene regia)* is rare in Indiana and is known to grow only at this and a few other nearby sites. Related to the more common Fire Pink (Plate 28), Royal Catchfly has larger flowers that are a brighter shade of red. The five slender petals are untoothed or only slightly toothed. As with other members of this genus, the calyx is sticky and often traps small insects on its surface, thus the common name, Catchfly. The bright showy flowers are borne in rather long, narrow, terminal panicles. The two to four foot stalk is unbranched and bears paired, sessile leaves that are lance-shaped and rounded at the base.

Culver's Root *(Veronicastrum virginicum)* bears narrow, branched, spike-like clusters of small white flowers at the top of an unbranched main stem. The densely positioned tubular flowers each have two projecting stamens. Slender, dark green leaves are two to six inches long and have sharp-toothed margins. They are paired on the lower part of the plant and grow in whorls of three to seven on the upper part. The genus name means "like Veronica," and at two to seven feet in height, the plant does look like a huge, white Veronica. The root has been used in herbal medicine because of its powerful emetic and cathartic properties.

The whitish flowers of Wild Quinine *(Parthenium integrifolium)* are composed of a small button of disc flowers with five tiny ray flowers widely spaced around the circumference. The numerous flowers form flat-topped clusters at the top of two to three foot stems. The long-stalked, toothed basal leaves are very large and have a coarse, rough texture. Smaller, alternate leaves are found on the bloom stalk.

Because of the rarity of Royal Catchfly, we did not collect a specimen of that plant for studio use. Maryrose avoids working from photographs or slides whenever possible, so she sat on a fallen tombstone and made her drawing at the site.

Maryrose had been taken to the cemetery that morning by a botanist who was doing an inventory, but who had to leave by noon. I was to pick her up by three, and then we were to continue on north to Indiana Dunes State Park to camp for a few days. By the time I finished my work at home and got the station wagon packed for camping, I was two hours late. My arrival was fortuitously timed, however—Maryrose was glad enough to see me that she had forgotten her anger at being kept waiting. We stopped again at the site on our way home, where I made restitution for my tardiness by holding an umbrella for two hours to shade her work while she painted the crucial parts of her drawing.

Although at times uncomfortable, angry, or worried during her five solitary hours on the shadeless prairie, Maryrose found the experience essentially satisfying. The sound of the wind in the six-foot Blue-stem Prairie Grass was a sound she had never heard before. Where she had pictured the virgin prairie as an endless, featureless grassland, she realized that it was actually a riot of color where many species of wildflowers bloomed in colonies among the tall grasses.

After this experience, she says, she has greater appreciation of the conditions pioneers faced when they crossed or settled the prairie. Because of the height of the grasses, there would have been a feeling of total isolation. Children wandering out of earshot from their parents could be lost and perhaps never found. Prairies remain grasslands because they are fairly dry, and periodic grass fires caused by lightning prevent shrubs or trees from growing large enough to produce shade that would eliminate the grasses. She realized how frightening it would be to travel on the prairie and see smoke in the distance. In all, Maryrose appreciated her experience and was awed by the richness and the profusion of the natural beauties surrounding her. Nonetheless, she was not sorry that the square of virgin prairie she explored was only an acre in size, so that she could find an open vista quickly in whichever direction she walked.

July 21, 1986

Maryrose Wampler

Plate 60

Our search for the Purple Fringeless Orchid finally met with success! I first saw several specimens growing along the shore at Yellowwood Lake in Brown County. Maryrose was working in a different area at the time, however, and missed them. Later we found this Orchid at Shakamak State Park, where Maryrose was able to paint it.

Purple Fringeless Orchid is considered rare—nationally it is even more uncommon than the Purple Fringed Orchid. In Indiana, however, we have been unable to find any reports of the Purple Fringed Orchid, while the Fringeless form is fairly common in some localities. We located several nice specimens driving along park roads and many more in the meadow areas of the park.

Purple Fringeless Orchid *(Platanthera peramoena)* is also known as *Habenaria peramoena*. The plant grows from one to three and a half feet tall and has two to five lance-shaped leaves. The large lower leaves sheath the stalk, and stem leaves become smaller and bract-like on the upper part of the plant. The flower head consists of a raceme of fifteen to fifty phlox-pink to rose-purple flowers. The lower leaves are often withered by the time the flower head is fully in bloom.

It is difficult to understand how a plant can be so rare and yet tolerate such a varied habitat. The plants I saw at Yellowwood were growing in wet humus in a densely wooded area. At Shakamak, Maryrose saw ten specimens growing singly along the edge of the woods as she drove the park roads. These sites were only partially shaded, and most were in well-drained positions at the tops of banks. I found the specimen Maryrose painted in a fairly dry meadow in full sun. The only common element that I could recognize among these varied sites was the rare beauty of the flowers themselves—and the abundant presence of bloodthirsty mosquitoes.

Rose Pink or Rose Gentian, also called Bitter-bloom *(Sabatia angularis)* has a thick, four-angled stem with paired flowering branches. The egg-shaped, opposite leaves clasp the stem of the one to three foot plant. Occasionally the beautiful white form is found growing among the commoner pink flowering plants.

With so many of the prominent summer flowers being yellow, vibrant rose-pink colonies of Sabatia along the roadsides are a refreshing sight. A biennial, Rose Pinks tend to form colonies as a result of the small round seeds falling near the parent plants. While most of the near relatives of *Sabatia angularis* are swamp dwellers, Rose Pinks are fairly adaptable; they grow just as happily on well-drained hillsides and meadows, in full sun along roadsides, and in light shade along edges of woods as they do in wet, low areas.

Field Milkwort *(Polygala sanguinea)* is a small plant whose flower looks a little like a clover bloom. The sparse alternate leaves are lance-shaped. The roots of Milkwort have a wintergreen odor when crushed. The plant was used in herbal medicine to stimulate lactation in wet-nurses.

The Green Milkweed *(Asclepias hirtella)* has dense umbels of greenish flowers in the leaf axils on the upper part of the one to three foot plant. The flower umbels are held close to the stem. While the flower clusters appear green, close inspection shows the petals are maroon with narrow white margins and only the upper part of the flower is light green.

July 26, 1986

Plate 61

Specimens for Plate 61 were found at Shakamak State Park. Partridge Pea *(Cassia fasciculata)* is a slender plant growing one to two feet tall. The bright yellow flowers are one to one and one-half inches in width and rise on separate stalks from the upper leaf axils. The flat flowers are composed of five broad petals of unequal size and have six conspicuous, dark purple anthers that droop from the flower's center. The alternate, compound leaves have twenty to thirty small, oval leaflets and are somewhat sensitive to the touch. The leaves fold at night and open flat again each morning. This annual plant produces flat seedpods that contain a row of seeds relished by partridges, quail, and other wild birds.

Early Goldenrod *(Solidago juncea)* is the first of the multitudinous members of the Solidago genus to bloom in Indiana. The open, gracefully arched plumes of golden yellow flowers are elm-like in shape. The plant grows one and a half to four feet tall and has large, toothed leaves that taper to long, margined stalks. In the axils of the slim, toothless upper leaves are tiny, wing-like leaflets. Both Partridge Pea and Early Goldenrod prefer well-drained habitats and are often found growing along roadsides at the top of banks.

Purple-headed Sneezeweed *(Helenium nudiflorum)* has showy, yellow-rayed, daisy-like flowers one to two inches across. The downward slanted petals are wedge-shaped and have three-lobed tips. The prominent disc is spherical in shape and dark purplish brown in color. The alternate, lance-shaped leaves are toothed and their bases form winged extensions down the stem of the plant. This plant does not produce wind-borne pollen, so that "sneeze" in the common name is not related to allergic reactions. In the past dried leaves were powdered and made into snuff. When inhaled, the powder caused sneezing.

Seedbox *(Ludwigia alternifolia)* looks something like a small Evening Primrose. The one-half inch yellow flowers have four petals backed by four prominent flat green sepals. Flowers are borne on short stalks in the leaf axils. The alternate lanceolate leaves are two to four inches long and are pointed at both ends. The smooth stem branches repeatedly to form a shrub-like plant two to four feet tall. The common name refers to the squarish, box-shaped seedpod. Both Seedbox and Sneezeweed like moist growing conditions and are often found along roadsides at the bottom of the drainage ditch.

The butterfly shown in this painting is a Tiger Swallowtail.

August 1, 1986

Maryrose Wampler

Plate 62

Specimens for Plate 62 were found at Potato Creek State Park. These plants are common in unmowed meadows and along fencerows and country roads. The small yellow five-petaled flower at left is Hybrid Loosestrife *(Lysimachia hybrida)*. Not actually a hybrid, but rather a separate species, this plant differs from Lance-leaved Loosestrife (Plate 45) in its stemmed leaves and lack of runners from the base of the plant. The long flower stems arise from the leaf axils and support one-half to one inch flowers that have toothed petals or petals abruptly pointed at the tip.

The large yellow daisy-like flowers in the center of the painting are Black-eyed Susans *(Rudbeckia hirta)*. Growing one to three feet tall, this rough, hairy plant has two to seven inch, three-veined leaves that are lance-shaped or oval. The flowers are two to three and a half inches wide and consist of bright yellow ray flowers surrounding a dark central cone (the black eye). A biennial, Black-eyed Susan forms a basal rosette of foliage the first year and blooms in its second year of growth. After blooming and producing seed the plant usually dies, although occasionally a plant will persist for more than one bloom season. These flowers are similar in appearance to the Prairie Cone Flower but have their petals held horizontally rather than drooping and have simple leaves rather than divided. In our experience Black-eyed Susans are much more widely distributed throughout the state than Prairie Cone Flowers.

At the top of the painting is Indian Hemp or Hemp Dogbane *(Apocynum cannabinum)*. Growing two to four feet tall, Indian Hemp bears terminal clusters of small greenish-white flowers. The smooth leaves are oval in shape and grow from opposite positions on the stem. A poisonous plant, Indian Hemp contains a milky-white sap and strong fibers in the stem. The fibers are the source of the common name, Hemp. Indian Hemp tends to grow in colonies and forms low thickets of dark green foliage spangled with white flowers.

One of the best-loved summer wildflowers, Butterfly Weed *(Asclepias tuberosa)*, is shown blooming at the right side of the painting. This brilliant flower is usually a burnt-orange color, although it ranges in orange shades from nearer yellow to nearer red. The complex flowers consist of five recurved petals surrounding an upraised, central crown. A milkweed, Butterfly Weed differs from other members of its genus in having alternate leaves and colorless sap. The two to five inch leaves are dark green in color and have a narrow lance-shape. Both stem and leaves have a rough, hairy texture.

The name Butterfly Weed may be related to the strong attraction this flower has for butterflies, or it may stem from the fact that this milkweed is the preferred host plant for the Monarch butterfly. A female Monarch is shown approaching the flowers and a zebra-striped Monarch caterpillar is shown feeding on a leaf. A gold-spangled green crysallis is at the left of the painting, hanging from the Indian Hemp leaf stem. When the caterpillars are ready to begin their metamorphosis, they often travel from their milkweed host. As the butterfly develops, the jade green chrysallis darkens to brown or black. When the butterfly emerges, it leaves the now-transparent case hanging empty.

Another common name for Butterfly Weed is Pleurisy Root. Chewing the tough roots has been used in herbal medicine as a cure for pleurisy.

August 1, 1985

Plate 63

It was in Monroe County east of Bloomington that we found these plants. They were growing in thickets on bottomland near a farmer's small private lake. On the left side of the painting is Wingstem *(Actinomeris alternifolia)*. This coarse, rugged plant can reach a height of eight feet and towers above competing foliage. The central stem branches repeatedly at the top and supports on slender stems the numerous one to one and a half inch yellow flowers. When the plant first begins blooming, the backswept petals give a graceful, floating aura to the expanse of blossoms. As spent flowers clutter the bloom head, however, the plant becomes progressively more unattractive. The common name refers to the rough leaves whose bases continue as wings down the stem of the plant.

Wild Senna *(Cassia hebecarpa)* bears clusters of yellow flowers at the top of the unbranched stem and in the upper leaf axils. The five-petaled flowers are about three-fourths inch wide and have ten stamens of unequal length tipped with prominent, dark brown anthers. The pinnately compound leaves are six to eight inches long and have five to ten pairs of one inch, oval leaflets. Wild Senna grows from three to six feet tall and may have several stems arising from the crown of the plant.

Woodland Sunflower *(Helianthus divaricatus)* is an attractive plant that grows in openings in the woods or along the edges of woodlands. It tends to form neat colonies of well-spaced, upright plants that do not branch, but may have a couple of side buds in the upper leaf axils in addition to the terminal flower. The blossoms are two to three inches across and have broad yellow petals or ray flowers surrounding a central disc. The paired, lance-shaped leaves are sessile or have stems that are under one-fourth inch in length. They are stiff and the upper surface is rough textured. The plant is smooth stemmed and grows from two to six feet tall.

Spotted Jewelweed *(Impatiens capensis)* bears orange flowers closely spotted with areas of darker color; in other respects it is similar to Pale Jewelweed (Plate 78). Another common name, Touch-me-not, refers to the ripe seedpods which coil like a spring when touched and scatter their seeds a considerable distance from the parent plant. Jewelweed is an annual that germinates fairly late in the spring and acts as a summer replacement for early spring wildflowers that have matured and died back by the onset of summer. The one inch flowers are composed of three colored sepals and three petals. Most of the visible flower is a large, sac-like sepal with a backward-curving spur. The upper part of the flower is made up of two smaller sepals. There are three small petals in the center of the flower, two of which are two-lobed. Spotted Jewelweed has thin, delicate, ovate leaves and juicy two to five foot branched stems that are nearly translucent. The juice is an effective treatment for poison ivy. If applied soon enough after exposure it can prevent the development of a rash, and it will alleviate the inflammation and itch in a rash that has already developed.

August 1, 1987

Plate 64

The three species shown in this painting are Swamp Milkweed, Hibiscus, and Pickerel Weed. All were found along the shores of lakes in Shakamak State Park.

Swamp Milkweed *(Asclepias incarnata)* grows abundantly over the entire state. Rivalling Butterfly Weed in beauty and prominence, it ornaments roadside ditches and other wet, low areas with colorful patches of dusty pink throughout July and August. The opposite lance-shaped leaves are about one half inch wide and four inches long. The species name, incarnata, meaning flesh-colored, refers to the color of the flower.

Monarch butterflies feed on members of the Milkweed family. We had previously found their larvae on Butterfly Weed. When we collected our specimen of Swamp Milkweed, there was a handsome young Monarch caterpillar about one and one-fourth inches long clinging tenaciously to the leaf on which he was feeding. We were impressed both by his voracious appetite and by the rapidity of his growth. During the two days Maryrose took to do the preliminary work for this painting, the caterpillar ate so much of her specimen that we had to collect a fresh one; he grew so large that she drew only his front part peeking over a leaf—fearing a full length portrait would be too obtrusive and dominate the flowers in the painting. Since our friend appeared to be approaching his mid-life crisis, we placed his mostly-eaten milkweed specimen securely among the branches of a vigorous plant to begin his metamorphosis in privacy.

Swamp Rose-mallow *(Hibiscus palustris)* is a showy herbaceous plant growing from three to eight feet tall. The flowers are four to seven inches wide and vary in color from white to almost red. A variety, *H. palustris forma peckii* (Plate 51)—considered by some botanists as a separate species—is white with a red or maroon throat. The graceful, five-headed pistil protrudes from the throat of the flower and is surrounded along the lower half of its column by yellow stamens.

This plant is usually found in wet areas along the shores of lakes, along rivers, or in drainage ditches along roads. It is fairly tolerant of habitat, however, and is frequently used as an ornamental accent in yards and gardens. The property manager at Shakamak rescued a large clump of mixed colors from a construction project by lifting it with a backhoe and depositing it in his backyard, where it now flourishes in multicolored splendor in relatively dry conditions.

Pickerelweed *(Pontederia cordata)* is an aquatic plant growing in shallow water from a rhizome buried in the mud. The thick, fleshy leaves are heart-shaped (thus the species name, cordata) and have spongy stems that float in water. The leaves themselves and the flower stalk with its one leaf rise a foot or more above the water level at bloom time. The flower stalk is topped by a spike of blue flowers about five inches long. Each flower has a conspicuous yellow spot on its upper petal. The flowers last only one day.

When we collected our specimen we were not familiar with the ephemeral character of Pickerelweed flowers. It was afternoon, and we were afraid we had missed peak bloom or had found a diseased specimen because of the withered appearance of the petal edges and the abundant fuzz that surrounded the blooms. Next morning, however, we had fresh blossoms all along the bloom spike and could see yesterday's flowers as tightly rolled cylinders withdrawing unobtrusively into the fuzz.

It is thought that the name Pickerelweed comes from the association of the habitat of the plant with that of the breeding grounds for pickerel fish. It is also possible that the name comes from medieval times in England when it was thought that plant shapes were related to similarly shaped animals, fish, or parts of the human body. When pickerel fish appeared in lakes or ponds where this plant grew, the similar long, slim shapes of fish and plant may have given rise to the belief that Pickerelweed would produce pickerel fish.

Pickerelweed is a natural source of food for both man and beast. In spring the tender shoots can be chopped and eaten raw as a salad ingredient or boiled as a vegetable. In fall the seeds can be gathered and eaten raw or roasted and ground into flour. Deer graze on the foliage during the summer, and ducks and other waterfowl feed on the seeds in late summer and fall.

August 4, 1986

Maryrose Wampler

Plate 65

All but one of the plants depicted here are aliens. They were found at Whitewater State Park. The light pink flower at the lower left is Bouncing Bet or Soapwort *(Saponaria officinalis)*. This one inch flower is composed of five backswept petals, shallowly notched at their tips. Occasionally there are extra petals, or the flowers may be double. These variations are believed to be associated with dry growing conditions. Color intensity also commonly varies, from almost white to dark pink. The opposite oval leaves are two to three inches long and have three to five conspicuous veins. This one to two and a half foot perennial from Europe is a vigorous, aggressive grower, spreading rapidly by means of underground stems and by the production of large amounts of seed. As shown in the painting, mature seed pods may be found on plants still producing fresh blossoms. Large drifts of multi-shaded pink seen along roads or in fields in the summer are usually Bouncing Bet. Both genus and common names are associated with the ability possessed by this plant to produce a soap-like lather. The Latin name, Saponaria, involves a word meaning soap. The common name Soapwort means soap plant, and Bouncing Bet is an old-fashioned name for a washer woman.

The spike of lavender blue flowers above Bouncing Bet is Tall Bellflower *(Campanula americana)*. The flowers of this beautiful native annual are flat and star-like rather than bell-shaped. The three-fourths to one inch flowers are light centered corollas that are divided into five pointed lobes. They range in color intensity from light to dark blue violet. Thin, lanceolate, toothed leaves are three to six inches long. In addition to the terminal bloom spike, solitary flowers or secondary bloom stalks may arise from leaf axils on the upper third of this two to six foot plant.

The showy blue flowers of Chicory *(Cichorium intybus)* are one and one-half inches wide and consist of square tipped ray flowers. Disc flowers are absent. Instead, dark blue anthers occupy the center of the flower. Before the one to four foot bloom stalk appears in late spring or early summer, the rosette of Chicory leaves looks much like a dandelion plant.

Chicory is another plant that forms large colonies. Spreading by means of seeds, it can turn the well-drained, gravelly soil along a roadside into a floral wonderland. But the wonder ends by noon. The short-lived flowers wither and lose their color, turning from breath-taking splendor to mundane drabness. When Maryrose was painting Chicory, we had to collect new specimens each morning, since we had found digging the plant a waste of time and effort because of the long, thick taproot. In subsequent years, we have had more Chicory than we wanted in our flower beds and lawn as a result of seeds blowing from the cut specimens we had placed outside the studio door.

Chicory came to America with the colonists from the Old World. They grew it in their gardens as a source of "greens," and in fall dug the roots and stored them in their cellars to produce blanched, head lettuce-like shoots during the winter. Roots were also roasted and ground to produce a coffee substitute or as a flavoring to be added to coffee. Chicory is still cultivated for these uses today.

The rounded flower heads of Red Clover *(Trifolium pratense)* are composed of closely crowded magenta pea-like flowers. The leaves of this familiar plant are divided into three oval leaflets. The leaflets are one-half to two inches long and have a lighter V-shaped pattern in the center. Introduced from Europe, Red Clover is used for hay and pasture and to improve soil texture and fertility. It is almost entirely dependent upon bumblebees for pollination.

The triangular-shaped, bright blue flower at the bottom of the painting is Asiatic Dayflower *(Commelina communis)*. The three-fourths inch flower protrudes on a short, slender stem from a folded heart-shaped bract. While each flower lasts only one day, the bract contains several buds. Two large, rounded blue petals form the top of the flower, and a small white petal lies behind the pistil and stamens at the bottom of the flower. Asiatic Dayflower grows to three feet in length, but tends to recline on the ground with only the branch tips standing erect. The three to five inch leaves are pointed at the tips. Their rounded base sheaths the stem. This annual plant is of Asian origin and in spite of its beauty is generally considered a weed. Spreading rapidly by seeds and rooting at the nodes wherever the stem touches the ground, Dayflower tends to crowd out more desirable plants in both the flower and vegetable garden.

August 5, 1984

Plate 66

Specimens for Plate 66 were found at McCormick's Creek State Park. On the left side of the painting is a small clump of Indian Pipes or Corpse Plants *(Monotropa uniflora)*. Entirely lacking in chlorophyll, this ghostly denizen of the deep forest has no green coloring and no need for light. Its nutritional requirements are met by the breaking down of organic forest litter by soil fungi. The flowers require pollination by insects, however, and produce seeds in the same manner as green plants.

Indian Pipes grow from three to ten inches tall and have a single, nodding flower at the top of each stem. Shaped like the bowl of a pipe, the one-half to one inch flowers are composed of four to six petals surrounding a broad, flat-topped pistil ringed with yellow anthers. The thick, succulent stems bear alternate scales or bracts in place of leaves. Although solitary plants are sometimes found, clusters of stems usually arise from a ball of matted, fibrous rootlets. The entire plant, with the exception of the yellow anthers, is usually a waxy white. It can be pinkish in color, however, and occasionally in the fall a deep rose shade is seen. The plant blackens as it ages and becomes tough and fibrous. Blackening also occurs where a fresh plant is bruised or damaged.

When the flower is pollinated and the seeds start to mature, the top of the stem straightens and holds the capsule erect. When ripe, the capsule splits along its sides and allows the wind to blow away the small seeds, starting new clusters of Indian Pipes where they find favorable conditions.

For many people, Indian Pipe is a plant of mysterious allurement. Perhaps this is because its strange shape is often seen glistening in places too shadowed for other plants to grow. While by no means rare, it is uncommon enough to be noted with interest and excitement when seen along a woodland trail. Indian Pipe has been used in the past as a medicinal herb by both Indians and early settlers. Sap from fresh plants was used as a treatment for sore eyes. Dried plants were used to brew a sedative that was used as a pain killer and to treat epileptic seizures.

Nodding Pogonia or Three-birds Orchid *(Triphora trianthophora)* is also a flower of the deep woods. This elusive plant is often found growing on rotting wood or in rich humus. After blooming, its underground tubers may lie dormant for several years before again sending up stems. Tubers send out short, slender stolons that form additional tubers at their tips. As a result, Nodding Pogonias tend to form small, tight clumps made up of several stalks.

The fragile stems are two to fifteen inches tall and have two to eight small, pointed ovate leaves that clasp the stem. The flowers arise from upper leaf axils. As buds the flowers are held in a nodding position, but their stem straightens as they expand so that the flowers are usually held erect. The one inch flowers may be pinkish or white. They are composed of three sepals and three petals. The upper sepal arches forward over the flower's lip and the two lateral sepals spread downward. The two upper petals are similar in color and texture to the sepals, but are shorter and form a hood above the lip. The rounded lip, formed by the third petal, is three-lobed and extends downward. Its edges are crinkled and it is marked by three green crests extending from the flower's throat. A single stem may bear from one to six flowers that somewhat resemble birds in flight. The usual number is three, thus the common name Three-birds Orchid. The withered flower often persists attached to the lower end of the pendulous seed capsule.

These flowers last only a day or so, and all tend to open at the same time, although a very few open before or after the main flush of bloom. When we were searching for specimens, we found a couple of stem clumps filled with buds but showing no open flowers. A few flowers had bloomed earlier and were developing into seed capsules. Over a period of two weeks, we kept checking the same clumps, unable to believe that there could be so many buds but no blossoms. Finally we hit the jackpot! Not only were almost all the buds blooming on the clumps we had spotted, but flowers were open all along the trail and in the surrounding woods where we had not suspected their presence. The plant is so small and delicate in appearance that it is difficult to see unless it is in bloom.

August 8, 1987

Plate 67

When deciding on exciting, unspoiled, wilderness areas for wildflower prospecting, we are inclined to think in terms of far-away places. We are sometimes surprised, therefore, by the richness and beauty of nearby woodlands, lakes, and streams. About a twenty-five-minute drive from our home, Yellowwood State Forest was a favorite fishing and relaxing spot of my college days. I can remember collecting my fishing gear, a good book, and a generous supply of junk food and going to Yellowwood Lake to escape the hassles of civilized life. There I would rent a rowboat and spend the day drifting along the shaded shores and inlets—reading, eating, and looking, but seldom catching any fish. Every now and then a fresh water jellyfish lazily pulsated its way through the water near my boat.

After an absence of thirty years, I was happy to return to Yellowwood this summer with a fellow wildflower enthusiast who knew the place where Purple Fringeless Orchid grows (Plate 60). While I was familiar with the lake itself, I had never done any hiking in the surrounding woods, and was amazed at the lush, almost rain-forest look of the area. In addition to the deciduous forest, there is a mature pine plantation where I saw leaves of Twayblade colonies that had bloomed in the spring and leaf rosettes of Rattlesnake Plantain that would bloom in the fall.

Most spectacular of all are the acres of American Lotus that have proliferated during the past few years, filling the shallow areas of the lake with their beauty and subtle perfume. American Lotus *(Nelumbo lutea)* has pale yellow flowers five to ten inches wide. Their long stalks hold them two to three feet above the surface of the water. The large round bowl-shaped leaves—a foot or more across—have their stalk attached in the center and are usually held above the water a foot or so below the level of the flowers. Each bloom is displayed in its circlet of leaves like a huge, solitary jewel in its setting.

American Lotus is a good source of wild food. The unexpanded leaves taken early in the season can be cooked as a green vegetable. The sweet potato-like tuber, though hard to collect because of its location in the muck under the shallow water, is delicious when baked.

The dense spikes of small magenta flowers shown at the right of the painting are Lythrum or Purple Loosestrife *(Lythrum salicaria).* The four to six petaled flowers are one-half to three-fourths inch wide and grow on plants four to six feet tall. The sessile, lance-shaped leaves are one and one-half to four inches long and grow in pairs or sometimes in whorls of three. While there are smaller species of Loosestrife that are native to Indiana, this showy perennial is an escaped garden flower from Europe. It fills acres of wet meadows and swamps in northern Indiana and ornaments ditches and lake shores in many other locations in the state, and is so invasive and rank in growth that it crowds out native water-loving plants that are important to wildlife. The temptation to plant it (it is available from most mail-order nurseries) should therefore be resisted.

August 11, 1986

Plate 68

Specimens for this painting were found in swampy areas near Hovey Lake in Point Township in Posey County. One of them, Meadow Beauty, is not considered threatened or endangered in Indiana, but it is not a common flower either. Maryrose had wanted to paint Meadow Beauty, but we had never seen it growing. The efforts we had made specifically to locate it had not met with success. We were delighted, therefore, to stumble across a previously unlisted site while looking for other plants.

At the left side of the painting is White Snakeroot *(Eupatorium rugosum)*. This plant is related to and similar in appearance to Blue Mistflower (Plate 71) and is sometimes called White Mistflower. The fluffy, snow-white flowers are each about one-fourth inch wide and form a many-flowered, flat-topped, terminal cluster. Paired, ovate leaves are long stemmed and sharply toothed. The plant grows two to four feet tall.

White Snakeroot likes moist growing conditions and is common along the edges of rich, lowland, wooded areas. It is frequently seen along lanes and country roads where they pass through woodlands. Cattle grazing on White Snakeroot produce milk that is dangerously toxic to humans. This is the plant blamed for causing the death of Nancy Hanks Lincoln.

The flowers of Sharp-winged Monkey-flower *(Mimulus alatus)* are indistinguishable from those of *M. ringens* (Plate 53). The characteristics that separate the two species are the length of the flower and leaf stems and the slightly winged stem of Sharp-winged Monkey-flower. M. alatus has stalked leaves and nearly stalkless flowers, while M. ringens has stalkless leaves and long stalked flowers.

Meadow Beauty or Deergrass *(Rhexia virginica)* has one to one and a half inch purplish-rose flowers composed of four broad, flat petals which form a lovely setting for the eight long, curved, golden anthers. The branched plant has square stems, slightly winged on the angles, that are one to two feet tall. The paired, ovate leaves are rounded at the base and have finely toothed margins. They are light green in color and have three to five prominent parallel veins.

Indian Tobacco *(Lobelia inflata)* is the most common of the Lobelias. The sparsely flowered plant bears tiny three-eighths inch flowers in shades of pale blue. The one to two foot plant is usually branched and produces its flowers singly in the leaf axils. Though small, the flowers have the characteristic three-lobed lower lip and are easily recognized as Lobelias. Alternate leaves, which are one to two and one-half inches long, are ovate and have toothed margins. This annual plant contains a narcotic poison, and dried leaves were used by Indians for chewing and smoking.

August 13, 1986

Maryrose Wampler

Plate 69

The flowers for this painting, Cardinal Flower *(Lobelia cardinalis)* and Tickseed Sunflower *(Bidens aristosa),* are found in bottomland and other low areas where moisture is abundant. Cardinal Flower has brilliant red blossoms that form a terminal spike six to twelve inches long on a two to four foot unbranched leafy stalk. Individual flowers are two lipped tubular corollas about one and one-half inches long. The upper lip is divided into two petal-like lobes and the lower into three. The stamens are united to form a tube that extends beyond the flower lobes. A narrow, leaflike bract is found beneath the green calyx of each flower. Alternate lanceolate six inch stem leaves are toothed. The flowering stalk arises from a basal rosette of leaves that sends out short stolons in late summer to form new plants that bloom the following year. Cardinal Flower is sometimes used as a garden flower, and several named varieties are available from commercial nurseries. Hummingbirds are important pollinators of Cardinal Flowers.

People who try to grow Cardinal Flower often experience failure because the plant has an unusually shallow root system. Although it can withstand cold temperatures, it must have a heavy mulch to protect it from frost heaving or drying during the winter months. In nature, this protection is provided by the heavy growth of neighboring grasses and plants that form a thatch as winter approaches. Since Cardinal Flower can grow happily in shallow aquatic conditions, the winter protection is sometimes a few inches of water.

Although Cardinal Flower is a perennial plant and produces seeds in abundance, it is seldom found covering large areas. Perhaps this is due to its special needs for adequate moisture and winter protection. While it is found in most sections of the state, it is not common. Usually it is seen growing in small colonies in roadside ditches, in alluvial soil bordering small streams, or along the shallow margins of ponds or lakes. One of our favorite experiences when on wildflower expeditions is encountering the bright red of Cardinal Flowers flashing against the dark green leaves of surrounding cattails.

Tickseed Sunflowers, also called Spanish Needles or Sticktights, have showy yellow daisy-like flowers one to two inches wide. The slender, leafy, much-branched stems are one to five feet tall. Opposite six inch leaves are pinnately divided into narrow, toothed segments. Both the genus and common names relate to the effectiveness of this plant's seed distributing mechanism. The flat seed has two teeth *(Bi-dens)* that are sharp as needles and stick tightly to anything they touch. Tickseed Sunflowers are widespread and grow in great numbers. In late summer, they can turn an entire valley floor into a lake of gold.

August 15, 1986

Plate 70

When walking along roadsides and through meadows and waste areas of Indiana in late summer, people often fail to notice the delicate beauty of flowers they pass. Perhaps that is because riotous growth at this time of the year crowds plants so. They strive for light and nourishment, and the eye doesn't differentiate plants and flowers as individuals. Or perhaps the observer's struggles against wild summer growth in lawn and garden have led to the bigoted notion that summer flowers are merely noxious weeds. Whatever the reason, it is a shame. There are many plants in bloom this time of year that deserve attention.

This painting shows five such. The first plant, beginning at the lower left and progressing clockwise, is Groundnut *(Apios americana)*. The flowers of this vine resemble those of pea vines. Maroon to chocolate in color, they bloom in tight clusters on stems three to five inches long. It is important to learn to recognize the five to seven broad, sharp-pointed leaflets of this plant because the flowers tend to be hidden beneath them. Blooming from July through September, the fragrant flowers of the Groundnut produce seeds that can be eaten like beans or peas. The name comes from the string of small tubers that comprise the root system. These tubers are edible and can be used in soups or fried like potatoes. Their mild, turnip-like flavor can best be enjoyed when they are served fresh and hot. If given room to run, the plant can be used in the garden as an ornamental by planting the seed or the tubers.

Blue Lettuce *(Lactuca floridana)* is a beautiful, airy flower resembling a miniature chicory. The flowers are only one-half inch across and bloom from July to October in a head at the top of a three to seven foot stalk. The deeply cut leaves resemble those of dandelions. The leaves and stem contain a bitter white sap. Young leaves are used like wild greens, but need to be cooked in one change of water to lessen the bitter taste.

Hollow Joe-Pye-Weed *(Eupatorium fistulosum)* displays a domed head of soft, fluffy pink blossoms on stems two to seven feet tall from July through September. The long, pointed, rough leaves are found in whorls of four to seven (usually six). This plant is believed to be named after an Indian named Joe Pye, who supposedly used it as a medicinal herb to calm fever. Commonly called Maids of the Meadow in southern Indiana because of its growth habit, it is extremely attractive to butterflies: a broad panicle of flowers may support as many as a dozen large butterflies hanging in apparent intoxication (pictured above, a Red Spotted Purple; below, a Pipevine Swallowtail).

Hedge Bindweed *(Convolvulus sepium)* belongs to the Morning Glory family. The flowers are usually about two inches across and may be either white or pink—sometimes white flushed pink. The leaves are two to five inches long and are arrowhead-shaped. Bindweed grows in thickets and along roadsides. It is especially partial to neglected gardens and flowerbeds. Those sensitive to informal, natural beauty might prefer it to the cultivated garden flowers upon which it encroaches.

Lady's Thumb *(Polygonum persicaria)* is sometimes called Redleg because of the color of the lower stems. One of several species of Smartweed, Lady's Thumb has knotted stem joints with a papery sheath at each joint. The plant grows six to twenty-four inches tall and blooms June through October. The Smartweeds belong to the Buckwheat family and have a head of tiny, bell-shaped pink flowers. The name Lady's Thumb comes from a dark green, thumbprint-shaped area in the center of the leaves. This trait is highly variable in both size and intensity and tends to be stronger on the lower leaves of the plant. Young leaves of Lady's Thumb are used as wild greens.

August 15, 1983

Maryrose Wampler

Plate 71

On the left side of Plate 71 are the flowers of Biennial Gaura *(Gaura biennis).* The long-tubed, four-petaled flowers are white when fresh but become a blush pink as they age, so two colors of flowers are often found on the same plant. The long, slim petals are all on one side of the flower and eight showy stamens arch from the other, giving the blossom an unusual asymmetrical appearance. The stigma has the cross-shape typical of the Primrose family of which this plant is a member. The blossoms open two or three at a time from clusters of buds on long wand-like spikes. The plant is bushy and grows two to five feet tall. Because of its height and slender stems, the plant is often found arching gracefully rather than growing stiffly upright. This airy wildflower is unobtrusive and easily overlooked, but the unusual grace and beauty of its slender flowers are worth a careful search and close inspection.

Blue Mistflower *(Eupatorium coelestinum)* is the only blue member of the large Eupatorium genus. Numerous one-fourth inch flowers form dense, flat-topped clusters. Composed of disc flowers, the blossom clusters have a fuzzy appearance because of the many protruding stamens. Mistflower grows one to three feet tall and has paired, arrow-shaped leaves. It is often found in dense colonies which make beautiful spots of powder-blue in the summer landscape. Because it is a perennial and similar in appearance to cultivated Ageratum, Mistflower is sometimes called Hardy Ageratum.

The flowers of Roundheaded Bush Clover *(Lespedeza capitata)* are found in dense, bristly, rounded clusters. The flowers are creamy white and often have a pinkish base. The two to five foot unbranched stems are crowded with short-stemmed, clover-like leaves composed of three oblong leaflets. Both stem and leaves are covered with soft, silvery hairs. Small oval seedpods form, each of which contains a single seed. After frost this plant turns a chocolate brown and is conspicuous throughout the winter.

False Foxglove *(Aureolaria flava)* has yellow bell-shaped corollas an inch across with five wide-spreading lobes. The smooth, gray-green stems are three to six feet long and often arch so that the flower-bearing tips are near the ground. The large leaves near the base of the plant are pinnately lobed. Leaves near the top of the plant are small and unlobed. False Foxglove is believed to be partially parasitic on the roots of oak trees. This plant is sometimes assigned to the genus Gerardia instead of Aureolaria.

Found at sites in Spring Mill State Park, the plants in this grouping like loose soil with sharp drainage such as that provided by prairie conditions or by raised gravelly roadsides. These plants, particularly the False Foxglove, can also tolerate light shade. These conditions are produced where roads pass through forested areas or in the zone where prairies blend into woodland.

August 17, 1986

Maryrose Wampler

Plate 72

The specimens for Plate 72 were selected on the basis of their unusual height. The tallest of the four species shown is Tall Coreopsis *(Coreopsis tripteris).* This astounding plant grows as high as ten feet, bearing hundreds of one to one and one-half inch yellow daisy-like flowers on the top few feet of its stem. Tall Coreopsis is a persistent perennial and makes a spectacular ornamental for gardeners who need flowers ten feet tall.

When the flowers first open, the disc flowers are tawny in color and are surrounded by eight broad yellow ray flowers (petals) which are rounded at the tip. As the flower matures, the center becomes dark brown, first as a dark ring around the golden center and finally as a completely dark center. Smooth stalked opposite leaves are divided into three lanceolate segments two to four inches long. The leaves are similar in appearance to those of Bush Clover. An anise-like fragrance is noticeable when any part of the plant is bruised.

Giant Hyssop *(Agastache scrophulariaefolia)* has spikes of purplish or light colored flowers at the tips of the branched stems. Tubular flowers are crowded on the spikes and are intermixed with purplish bracts. The four protruding stamens contributed by each flower give a fuzzy appearance to the floral spike. The long-stalked leaves are coarsely toothed and have a thick coating of whitish hairs on their undersurface. While the height of this plant is usually given as two to five feet, this specimen was closer to eight feet in height.

The clean yellow flowers of Common Evening Primrose *(Oenothera biennis)* are to be seen along almost any roadside in late summer. The one to two inch flowers are composed of four broad petals that open wide at twilight and wilt the next day when the sun gets hot. Blossoms arise from the tops of long floral tubes that drop when the seedpod begins to develop. The flowers have eight golden stamens and a prominent *X*-shaped stigma. The numerous alternate leaves are four to eight inches long and are rough-textured and hairy. One to six foot stems often branch toward the top and sometimes show a reddish tinge of color.

As the species name indicates, Evening Primrose is a biennial and usually blooms in its second year of growth. After blooming the plant dies, but remains standing to disperse its seeds during the winter and following spring. When seed capsules mature, the segments reflex and produce a stem of attractive gray flower-like structures useful in dried arrangements. Young leaves and roots of first-year plants are edible and are especially good in the spring or fall when their flavor is mild.

Wand-like Bush Clover *(Lespedeza intermedia)* bears dense clusters of showy red-violet pea-like flowers, mostly at the tips of the branched stems. The erect, shrub-like plant grows two to four feet tall. The characteristic clover leaflets are small and oval in shape. The leaves are fairly sparse and do not obscure the graceful, wand-like appearance of the stems.

August 21, 1986

Plate 73

The specimens shown in Plate 73 were found in the vicinity of Potato Creek State Park near La Porte in northern Indiana. The three spikes of purple flowers are three separate species of Liatris. Those that have flower heads spaced at intervals along the bloom spike are commonly called Blazing Stars, while those that have flower heads closely enough spaced to form an unbroken spike of bloom are called Gayfeathers. On the left is *Liatris scariosa.* This Liatris is characterized by large hemispherical flower heads on stalks whose length is two or more times the width of the flower head. Flower heads are usually about an inch wide and contain thirty to one hundred disc flowers. The backs of the flower heads are covered with outflaring, broadly rounded bracts which may be purple or purple edged. Leaves are lanceolate, tapering to the base of the stem, and are about eight inches long at the bottom of the one to four foot plant. Upper leaves become progressively smaller.

The middle spike of Blazing Star is *Liatris aspera* or Rough Blazing Star. This species is similar to L. scariosa except that the flower heads are short stemmed or stemless and have fewer disc flowers. These differences produce a slimmer spike of smaller flowers.

The third spike is *Liatris spicata,* commonly called Dense Blazing Star or Gayfeather. The small crowded flower heads are sessile and have only five to nine disc flowers each. The buds occupy the upper four to twelve inches of the stem and, as with all Liatris species, bloom from the top downward. Numerous grass-like leaves, as much as twelve inches in length, form a clump at the base of the plant. The flowering stem is also leafy, but stem leaves decrease in size as they ascend the stem. Gayfeather favors a wetter habitat than the other two Liatris species.

The spike-like clusters of pinkish-violet flowers on the right are False Dragonhead or Obedient-plant *(Physostegia virginiana).* The three-fourths to one inch flowers form clusters four to eight inches long on the upper part of stems that often branch near the top of the plant. Flowers are two lipped corollas with the upper lip arching hood-like over the purple-spotted lower lip. The common name Obedient-plant reflects the fact that if the vertical rows of flowers are moved to one side, they will remain there for a time. The opposite lanceolate leaves are narrow and pointed at the tips. They have sharp, incurving teeth along their margins. Leaves may be four inches long at the base of the plant and become smaller toward the top. Physostegia is a common, old-fashioned garden plant that is able to grow and bloom with little care. While varieties selected for garden use are usually about two feet tall, wild plants often reach as much as four feet in height. Both pink and white clones are available as garden plants.

At the bottom of the painting is Flat-topped White Aster *(Aster umbellatus).* These small flowers have only ten to fifteen rays that tend to curve downward. The yellow disc turns purplish with age. Toothless, lance-shaped leaves may be as much as six inches long and taper at both ends. The plant grows from two to seven feet tall and blooms earlier in the season than most other asters.

Whenever Maryrose decides to include an insect in a painting, suitable specimens seem to disappear from sight. Occasionally, however, one will virtually flaunt itself, apparently insisting on being used as a model. The Pandora Sphinx caterpillar in the foreground was such a case. Maryrose first encountered it when she reached outside one morning to unzip her tent door and found the creature clinging to the zipper tab. She relocated it gently but firmly in some nearby bushes, but the next day it again presented itself, posing at eye level above the drawing board. Even though its color wasn't quite what she wanted in the painting, she decided that such persistence deserved to be rewarded.

August 22, 1986

Maryrose Wampler

Plate 74

The plants shown in Plate 74 require dry, acid growing conditions such as those found in upland oak or coniferous forests. Specimens of Downy Rattlesnake Plantain *(Goodyera pubescens)* and Ground-cedar *(Lycopodium flabelliforme)* were found in an old section of a Christmas tree farm in Fulton County where a group of pines had been allowed to grow to maturity.

The tiny Lady's Tresses came from an old pine plantation near Griffy Lake in Monroe County. Maryrose had been lamenting to a friend who was knowledgeable about wildflowers that she had been able to find only two species of plants that were suited to this painting. Our helpful friend had been particularly interested in the towering Lady's Tresses we had found for Plate 47 and told us that he had miniatures growing in the pines behind his house. The specimen he brought us was *Spiranthes tuberosa,* listed as an endangered plant in Indiana. There were over thirty specimens at this previously unlisted site! The one shown in the painting is now a permanent resident at the herbarium at Indiana University.

Downy Rattlesnake Plantain grows from six to twenty inches tall and bears a slender spike of small greenish-white flowers. A member of the Orchid family, the one-fourth inch flower is composed of a hood, made up of the upper sepal and two united petals, that arches over a cupped lip petal. The thin flower stem is topped by as many as eighty of these small flowers. Although the stem appears leafless, there are a few scale-like bracts present below the head of flowers. The rosettes of leaves are the most distinctive feature of this interesting plant. They form mats of dark blue-green, dramatically veined and netted with white. The leaves persist over the winter and are attractive throughout the year.

Tuberosa, the species name of Little Lady's Tresses, refers to the cylindrical tuber-like root whose top lies just beneath the layer of forest litter. In the spring two or three small, ovate leaves arise on short stalks from the top of the tuber. The leaves wither and disappear before flowering time. The bloom stalk is three to five inches tall and is very slender. It has a few tiny bracts along the lower part and bears minute white flowers in a single rank or spiral toward the top.

Ground-cedar is a clubmoss rather than a conifer and reproduces by means of spores released from candelabra-shaped cases. A single plant may cover several square yards with dense, shining green, cedar-like foliage. The six inch, upright stalks arise from long, horizontal stems that lie beneath the top layer of humus. This beautiful evergreen plant is popular for use in terrariums and was once commonly collected for decoration at Christmas time. Since manufactured Christmas ornaments are currently available in great number and variety, Ground-cedar is no longer gathered in quantity, and is again becoming fairly common.

August 26, 1986

Plate 75

The flowers in Plate 75 were found at Tippecanoe River State Park and are common summer roadside plants. In selecting these specimens, we discovered that Sneezeweed *(Helenium autumnale)* does not take kindly to being either dug or picked. Most plants that Maryrose has drawn and painted at their growing sites have been too rare to be disturbed. While this was certainly not true for Sneezeweed, she had to set up paints and drawing board along the side of the road for it regardless—all efforts to work on it in the studio ended with wilted flowers.

The general shape of Sneezeweed is similar to that of Purple-headed Sneezeweed (Plate 61) except that the plant is taller and the flowers are larger but not so numerous. The bright yellow flowers are daisy-like in shape and may be as much as two inches across. A spherical ocher-yellow center is surrounded by backswept petals that are wedge-shaped and have three prominent lobes at their broad tips. The winged stem is two to five feet tall and branches near the top. Alternate, lanceolate leaves are as much as six inches long and have toothed margins. In the past leaves of this genus were powdered to make snuff. When inhaled the powder caused sneezing.

The flower heads of Goldenrod are botanically grouped according to characteristic shapes. This painting shows an example of a spiked or wand-like Goldenrod *(Solidago species)*. Goldenrods dominate the landscape from midsummer until frost. An interesting activity for a summer day is to walk through a grassy meadow and observe the different colonies of Goldenrod growing there. When a Goldenrod seedling finds a favorable habitat and has enough vigor to compete with surrounding plants, it spreads vegetatively in all directions by means of underground stolons. An old colony may be several yards across and contain hundreds of plants, each of them identical to the original seedling. The colonies are often different species, and one colony may have flower heads shaped differently than those nearby. There may be small differences in color as well as form. People who experience allergy problems while engaged in activities such as this need to look outside the Solidago genus for the source of their problem; Goldenrods produce heavy, insect-borne pollen and do not contribute to the hay fever season.

The Ironweed shown in this painting is Missouri Ironweed *(Vernonia missurica)*. The plant is similar in appearance to Tall Ironweed (Plate 79) but has much larger flower heads. The flower head is about three-fourths inch wide and is composed of twice the number of disc flowers contained in a flower head of Tall Ironweed.

September 7, 1986

Marjorie Wampler

Plate 76

Specimens for Plate 76 were found at Pokagon State Park. This was the last of the paintings begun for this series. Needing eighty plates, Maryrose had saved one hoping to find Fringed Gentian, a flower that is becoming increasingly rare and difficult to locate. It needs wet, acid conditions where the ground water moves through the soil and does not become stagnant. A biennial, it may not bloom in exactly the same location each year. The plants she painted were found in a seepage area in sandy soil. Although the site had full exposure to sunlight, vegetation was sparse, perhaps because of the acidity of the soil. Fringed Gentian's natural range covers the entire state, but it is usually found in the sandy soils of the northern regions. Specimens growing naturally should never be disturbed. Picking reduces the plant's capacity to reseed itself, and digging is pointless because the plant dies after blooming. But gardeners with a sunny, wet meadow or seepage area on their property might profitably attempt to grow this spectacular wildflower from seed.

Apparently the site where these flowers were found usually has an abundant population of plants. We searched the area carefully, but without success. It had been very dry that year, and either seeds had not germinated or the young plants had not survived the drought. Just as we were about ready to give up, we found a single plant blooming in glory at the base of a small clump of trees. They were located where water rose to the surface and had been supplied with sufficient moisture even during the dry weather. Coming suddenly upon this one plant with its five freshly opened, vibrant blue blossoms, we felt like offering a prayer of praise and thanksgiving. At that point the huge flowers isolated against their setting of fresh green foliage were more beautiful to us than a whole field of them would have been. It was a sight we shall never forget.

The other flowers shown in the painting were found growing within a few feet of the Gentian. At the left is Purple Gerardia *(Gerardia purpurea)*. Red-violet flowers are about one inch long and have a five-lobed, inflated bell-like shape. They arise from the leaf axils and are almost without stems. The paired leaves are one to one and one-half inches long and are very narrow. This plant is an annual and grows one to three feet tall.

Spotted Joe-Pye-Weed *(Eupatorium maculatum)* is shown at the top of the painting. Purple stems, deep rose flowers, and flat-topped clusters distinguish this Joe-Pye from other species that grow in Indiana. This northern species has pointed, lance-shaped leaves that have a coarse, net-veined texture. Found in whorls of three to five, the leaves are three to eight inches long. The plant grows two to six feet tall.

The flowers of Fringed Gentian *(Gentiana crinita)* open in a swirl of four fringed petals that flare from the long corolla tube. The blossoms are about two inches long and open only when the sun shines. A long, pointed green calyx encases the lower part of the flower tube. Bright blue flowers are borne singly on long slim stems. The one to three foot plant is generously branched. Paired one to two inch leaves are rounded at the base and have a pointed tip. Smooth surfaced, they are a bright green color that emphasizes the blue of the flowers.

The flowers of Nodding Lady's Tresses *(Spiranthes cernua)* are positioned around the upper part of the six to twenty-four inch stem in a double spiral. The downward arching, fragrant white flowers are about one-half an inch long. Two side petals and the upper sepal unite to form a hood over the wavy-edged lower lip petal. The slender ten inch leaves die back before the flowers bloom.

Brook Lobelia *(Lobelia kalmii)* grows four to sixteen inches tall. White-centered blue flowers borne in loose racemes look large against the narrow leaves and thin, delicate stems. The three broad lobes of the lower lip dwarf the two small lobes of the upper lip. The plant does not look much like a Lobelia, but the flowers are easily recognizable.

September 8, 1987

Maryrose Wampler

Plate 77

Specimens for Plate 77 were found in an area north of Bloomington in Monroe County. Closed or Bottle Gentian *(Gentiana andrewsii)* and Turtlehead *(Chelone glabra)* were plants we had read about and knew we should be able to find. We had never located either one until we were taken to a site in Fulton County where Closed Gentian was known to grow. We found only one plant; while we were thrilled by its color and beauty, we did not want to disturb a single specimen. Nor did we have time to bring in materials to paint it at the site.

On our way back to Bloomington, we decided to check a side road where we knew we could collect a Bidens specimen and where we had seen Cardinal Flower several years before (Plate 69). We drove past the area we had wanted to inspect and were happy to find the Cardinal Flower still thriving. As we drove back to the highway after finding a place to turn around, we noticed a motorist behind us obviously impatient with our leisurely pace. Spotting a wide place in the road, I pulled over to let the car pass and was startled to hear sputterings and inarticulate gasps from Maryrose. When she recovered enough to assure me that I was not about to ruin a tire or ditch the car, I found that her excitement was caused by sighting a bank beside the pull-off that was blue with hundreds of gorgeous Closed Gentian blossoms. A short distance down the road, we were again thrilled to see spikes of Turtlehead that we had missed on our way by the first time. After traveling the state hoping to find these two beautiful wildflowers, here they were growing in profusion only a few miles from home!

The unusual flowers of Closed Gentian have deep blue petals that stay closed and are joined at the top by a fringed, whitish band or membrane. A tight cluster of several upright blossoms is at the top of the unbranched stem and a few flowers are often located in the upper leaf axils. A whorl of several leaves is found below the terminal flower cluster. The lower leaves are paired. Leaves are parallel-veined and ovate in shape. They may be as much as four inches in length.

Gentians are distinguished by flowers of beautiful, unusual shapes and by their intense blue color. Members of this genus tend to have exacting habitat requirements and are sparsely distributed in nature and exceedingly difficult to grow under cultivation. Many are biennials and require constant reestablishment if they are to be maintained in a garden setting. Closed Gentian is a happy exception to these generalities, however. A persistent perennial, it adapts readily to cultivation and provides a beautiful spot of fall color in either sun or shade if sufficient moisture is provided.

The flowers of Turtlehead are found in tight clusters at the tips of the stems. The turtlehead-shaped, two-lipped blossoms are white sometimes tinged with pink. The upper lip arches over the three lobes of the lower lip. The center lobe of the lower lip is shorter than the two outer lobes and adds to the appearance of a turtle's mouth. The opposite, lanceolate leaves are sharply toothed and have pointed tips. The plant is square-stemmed and grows one to four feet tall.

There is a rose-colored form of Turtlehead that is sometimes found escaped from cultivation. The native white-flowered plant also makes a good garden subject if enough water can be provided.

In the background of Plate 77, Drummer, a friend of our son's, is combing the tall grasses for gamebirds. The covered bridge is characteristic of many such historical structures still in use in Indiana. Maryrose added it to the painting because in fact there used to be a covered bridge near this site. Unfortunately, it burned a few years ago and was replaced by a modern bridge that is more functional but not nearly so picturesque.

September 9, 1986

Plate 78

The plants shown in Plate 78 can be found in wet areas at Lincoln State Park. The spike of flowers on the left is Blue or Great Lobelia *(Lobelia siphilitica)*. This showy Lobelia grows one to four feet tall and has stout, leafy stems. The dark blue flowers bloom at the axils of leafy bracts on the upper parts of the plant. Individual flowers are two-lipped. The upper lip has two lobes, and the longer lower lip has three lobes subtly striped with white. The two to six inch leaves are oval to lance-shaped and taper to a sessile base. In the past, Blue Lobelia was considered an herbal cure for syphilis, thus the species name. Blue Lobelia is commonly found growing along stream banks and in drainage ditches along wooded country roads. A well-grown spike of it is a memorable sight.

The flower at the top of the painting is Purple-stemmed Aster or Swamp Aster *(Aster puniceus)*. The narrow leaves of this three to eight foot plant are rough in texture and up to six inches in length. They taper to a narrow base that clasps the stem. The one to one and one-half inch lavender flowers have twenty to forty narrow rays and bloom in large, showy heads. While flowers of Purple-stemmed Aster are as beautiful as those of many cultivated varieties, the swampy growing conditions it requires and its robust size make this plant inappropriate for most flower gardens.

Boneset or Common Thoroughwort *(Eupatorium perfoliatum)*—an older spelling is Throughwort—is a hairy plant with dense, flat-topped heads of small, dull white flowers. Perhaps its most distinctive feature is its leaves. Joined at the base so that they completely surround the stem, the two opposite leaves give the appearance of a single leaf pierced in the center by the stem. The species name, *perfoliatum,* describes this characteristic, which has also given the plant its common name, Boneset. Based upon the Doctrine of Signatures, early herbalists considered the stem of this plant, "broken" by its leaves, to have its human parallel in broken bones. Boneset leaves therefore were incorporated into wrappings for broken bones to aid in healing.

The yellow flower in the foreground is Pale Jewelweed or Pale Touch-me-not *(Impatiens pallida)*. The one inch flower consists of three petals, two of which are lobed, and three sepals. The lower sepal is colored like the petals and comprises most of the flower. The other two sepals are green in color and smaller in size. The pale green, toothed, oval leaves are one and a half to three and a half inches in length. The seeds are contained in a swollen capsule that "explodes" when touched, scattering seeds in all directions. These active, entertaining seed capsules are the source of the common name Touch-me-not. Children delight in inviting their uninitiated friends to pick these pods and then watching their startled reactions when the ripe pod twists and pops open in their hand. The succulent, translucent stems and thin leaves of this prolific annual plant contain a juice that is effective in treating poison-ivy rash. I have found from personal experience that this is one herbal remedy that is truly effective. Bruised plant parts used as a poultice or juice painted on the affected area reduce the redness and swelling and eliminate the itching. The treatment needs to be repeated every few hours until the allergic reaction is spent.

September 21, 1983

Plate 79

Specimens for Plate 79 were found in a bottomland thicket in Monroe County. At lower left is Jimsonweed or Thorn Apple *(Datura stramonium)*. For many, particularly farmers, the truly spectacular beauty of this plant is obscured by its obnoxious characteristics. Growing one to five feet tall, this rank-smelling, extremely poisonous interloper from tropical America sprouts in quantity in abandoned barnyards, waste areas, and ploughed fields. The glossy, ovate leaves are two to eight inches long and are irregularly lobed. The seeds are produced in a two inch egg-shaped capsule covered with sharp spines. The fragrant flower is a five-lobed, three to five inch, trumpet-shaped corolla that may vary in color from almost white to violet. A long green tubular-shaped calyx sheaths the base of the flower. Blossoms open at sunset, are pollinated by night-flying moths, and wither the next morning. Apparently Jimsonweed was an early immigrant to the United States since "Jimson" was originally "Jamestown." The common name, Thorn Apple, refers to the shape of the spiny seed capsule.

The small purple flowers at upper left belong to Tall Ironweed *(Vernonia altissima)*. The individual flower is about one-fourth inch wide and is composed entirely of disc flowers. Ray flowers are absent. The alternate, sessile, lance-shaped leaves are six to ten inches long and are sharply toothed. The undersurfaces of the leaves are downy. The common name Ironweed refers to the hardness of the plant's stem. Since cows do not eat Ironweed, the plant thrives in pastures. As they graze, the cows in effect cultivate this weed by eliminating its competition. When small family farms were thriving, a common sight was a farmer walking through his pasture land using a sharp hoe on the Ironweed.

At upper right is the biennial Tall Thistle *(Cirsium altissimum)*. Growing three to twelve feet tall, this thistle differs from others of its genus in having leaves that are not deeply lobed or incised. The prickly-edged, lance-shaped leaves taper at each end and have a white, woolly undersurface. The ball-shaped violet flower heads are one and a half to two inches wide.

The small bluish flower at lower right is Heart-leaved Aster or Common Blue Wood Aster *(Aster cordifolius)*. The lower leaves, which are not shown in the painting, are two to six inches long and broadly heart-shaped. The one to four foot plants bear dense clusters of five-eighths inch, pale blue-violet flowers.

In the foreground is the Common Elderberry *(Sambucus canadensis)*. This three to ten foot shrub bears many flat-topped clusters of tiny, white, fragrant flowers in late spring and early summer. Pinnately compound leaves are four to eleven inches long and have five to twenty-two toothed, oval leaflets. In the fall, clusters of purplish-black berry-like fruit hang heavy ready for the enjoyment of animal, bird, or person. Elderberries commonly grow along fencerows where birds that had eaten the fruit deposit seeds in their droppings as they sit on the fence wire singing their appreciation.

Many people value Elderberries for making jelly, pies, or wine. While the fresh fruit is unpleasantly bitter, cooked or dried fruit is quite tasty. Elderflowers are also prized by people who enjoy natural foods. Small sections of Elderflowers are dipped in batter and deep fat fried. Dusted with powdered sugar or served with hot syrup, they have a delicious vanilla-like flavor.

September 22, 1983

Plate 80

Plate 80 shows six species of plants: Late Goldenrod *(Solidago gigantea),* Bittersweet *(Celastrus scandens),* Jerusalem Artichoke *(Helianthus tuberosus),* Evening Lychnis *(Lychnis alba),* White Fall Aster *(Aster pilosus),* and Ivy-leaved Morning-glory *(Ipomoea hederacea).* The specimens were found at Shades State Park.

Goldenrods are difficult for the amateur to positively identify. There are over a hundred species, and several of these may hybridize. Since paintings cannot be subjected to meticulous, analytical scrutiny we have chosen the species name *gigantea* for this specimen because it conforms generally to descriptions of that species. Because of the August to October bloom season Late Goldenrod is the common name.

Growing two to seven feet tall, Late Goldenrod has an attractive, plume-like flower head atop a smooth pale-green or purplish stem that often shows a whitish coating. Narrow, lance-shaped leaves are usually toothed and grow alternately on the stem. This plant prefers moist thickets or weedy fields.

Goldenrod is a persistent perennial plant that grows either from seeds or from underground rhizomes. Seeds germinating in spring produce a single flowerless stem during the first year of growth. The second year the plant blooms and sends out four or five rhizomes which bloom and spread their rhizomes the third year. This process may continue for decades, forming large, dense colonies that are all clones of the original plant. The flower stalks die in winter, but remain standing to disperse their seeds. Goldenrod is often wrongly accused of causing hay fever. The real culprit is Ragweed, which blooms at the same time. Ragweed's tiny green blooms go unnoticed, but release huge quantities of windborne pollen. The heavy, sticky pollen of Goldenrod is carried by insects rather than the wind. Goldenrod has been used as an herbal tea to soothe sore throats and colds. While Sweet Goldenrod is best known for this use, any Goldenrod may be tried. Young leaves and fully opened flowers produce an anise-flavored tea.

Bittersweet is a woody, twining vine that blooms in May and June (see Plate 26) in terminal clusters of tiny green flowers so inconspicuous that they are seldom noticed. In fall clusters of yellow-orange jacketed berries open to expose scarlet seeds, producing a dramatic, bicolored display. Bunches of Bittersweet are often sold in the fall at roadside stands.

Jerusalem Artichoke is a five to ten foot plant with a spectacular head of two to three and a half inch yellow, daisy-like flowers. The thick, rough leaves are four to ten inches long and may be oval or lance-shaped. Leaves on the lower part of the stalk are opposite, while those on the upper part are alternate. The plant thrives in such moist places as ditches, along streams, and in bottomlands. This plant is not related to artichokes, nor does it have any association with Jerusalem. The common name is a corruption of the Italian word for sunflower, *girasole.* A native sunflower, it was cultivated by the Indians for its edible tubers. These also provide valuable food for herbivorous animals. The tubers may be boiled or baked like potatoes or eaten raw in salads. Tubers are formed in late summer and fall as the days become shorter.

The interesting flowers of the Evening Lychnis resemble blossoms on a balloon. The one inch white or pinkish blooms have five deeply notched petals. The calyx is finely veined and more or less inflated. This plant produces male and female flowers on different plants. The calyx of the male flower is more slender and has fewer veins than the female flower. The female flower may also be recognized by the female parts which extend out beyond the rather flat face of the flower. Evening Lychnis is pollinated by night-flying moths, which are attracted by the white flowers and by the strong, sweet fragrance.

White or Frost Asters are among the last flowers of the season to bloom. Seen from September into November, these perennial plants often flower after the frosts have killed other plants. Like Goldenrod, Asters remain standing during the winter, dispensing their windborne seed. Since Asters seed themselves almost as readily as annuals, they are often found in large numbers on disturbed ground. There are numerous species of Asters that are difficult to differentiate.

Ivy-leaved Morning-glory, so named because of the three-lobed shape of the leaf, may be white, pink, or blue. The vine spreads several feet and may be invasive in fields and along roadsides. Morning-glory is an annual vine introduced from tropical America.

September 30, 1983

GLOSSARY

Alien: Foreign; native to a continent other than North America.
Alluvial soil: Sandy or silty soil deposited by flowing water on the valley floors of streams or rivers.
Alternate leaves: Leaves that occur singly along the stem; not opposite or whorled.
Annual: Plants that germinate, produce seed and die in a single growing season.
Anthers: The structure at the tip of the stamen that bears the pollen.
Axil: The crotch formed where a leaf joins the stem.
Basal rosette: A circular cluster of leaves found at the base of a plant.
Biennial: A plant that requires two years to complete its life cycle. Usually a basal rosette of leaves is produced the first year that lives through the winter. A bloomstalk and seeds are produced the second year and the plant dies.
Bracts: Modified leaves, usually small, that are often found at the base of the flowers.
Calyx: A collective term for the sepals of a flower.
Chlorophyll: The green coloring matter of plants which, in the presence of sunlight, produces carbohydrates from carbon dioxide and water.
Clone: Plants that are vegetatively reproduced from a single individual. Such plants may result naturally from stolons or side shoots around a parent crown or artificially by budding or grafting.
Compound leaves: Leaves that are divided into several separate leaflets.
Corm: A short swollen underground stem resembling a bulb in appearance.
Corolla: A collective term for the petals of a flower.
Deciduous: Woody stemmed plants whose leaves drop in autumn; not evergreen.
Disc flowers: The tiny tubular flowers crowded into the center of daisy-like flower heads.
Generic name: The first of the two scientific names by which plants are identified. A plant's scientific name is the same all over the world although it may have many different common names in different countries and languages.
Herbaceous plants: Non-woody plants whose stems die back to the ground each winter.
Hips: The apple-like fruits of roses which contain the seeds.
Lanceolate leaves: Shaped like a lance; several times longer than wide, tapering at both ends but broadest below the middle.
Native: Indigenous to North America; not introduced from other continents by man.
Naturalized: Alien plants that are growing in the wild without cultivation or protection.
Opposite leaves: Leaves that are situated in pairs on the stem.
Ovate leaves: Broadly rounded leaves, elliptical or egg-shaped with the widest part near the base.
Palmate leaves: Compound leaves where the leaflets are arranged like the fingers of a hand.
Panicle: An elongated, branched flower cluster.
Perennial: A plant whose roots and crown live through the winter and sprout anew each year.
Pedicel: The stem of an individual flower in a head of blooms.
Petiole: The stem of a leaf.
Pinnate leaf: A compound leaf where the leaflets are arranged in pairs or alternately along a central stalk.
Pistil: The female part of a flower composed of an ovary, style and stigma.
Pollen: The male spores produced by the anther.
Pollinator: The agent that transfers pollen from plant to plant usually thought of as an insect; however, wind is the main pollinator for many plants, and hummingbirds carry pollen when they gather nectar.
Raceme: A spike of flowers arranged singly along a stalk where each flower has its own stem.
Ray flower: The petal-like flowers surrounding the central disk of daisy-like flower heads.
Rhizome: A thick food-storing underground stem usually found in a horizontal position.
Sepals: Segments of the calyx, outer covering of the flower bud. Sepals are usually green, although in some flowers, they are colored like petals.

Sessile: Stalkless, usually refers to leaves that grow from the stem without a petiole but may refer also to flowers that are nestled stemless in leaf axils.

Spadix: The club-shaped structure characteristic of members of the Arum Family that bears tiny flowers along its surface.

Spathe: A large bract that hoods or enfolds a flower cluster as in members of the Arum Family.

Species name: The second of the two scientific (Latin) names by which plants are identified. A plant's scientific name is the same all over the world, although it has many different common names.

Stamen: The male element of a flower, usually numerous the stamen consists of a pollen-bearing anther supported at the tip of a slender stem called the filament.

Stigma: The sticky tip of the pistil that receives the pollen.

Stipules: A leaf-like growth, usually small, at the base of a leaf stalk.

Stolon: A horizontal stem growing from the crown of a mature plant that roots at the nodes and produces new plants identical to the parent plant.

Style: The part of the pistil that connects the pollen-receiving stigma with the seed-producing ovary.

Tuber: A fleshy underground stem bearing buds that can vegetatively reproduce new plants.

Umbel: An umbrella-like cluster of flowers where all the stems radiate from a common point.

Whorl: A cluster of three or more leaves or flowers radiating from a common point on the stem.

Winged stems: A thin leaflike membrane extending along a stem or other plant part.

BIBLIOGRAPHY

Aldrich, James R., John A. Bacone, and Michael A. Homoya. 1986. List of Extirpated, Endangered, Threatened, and Rare Vascular Plants in Indiana: An Update. Indianapolis: Indiana Department of Natural Resources.

Birdseye, Clarence, and Eleanor G. Birdseye. 1972. Growing Woodland Plants. New York: Dover Publications, Inc.

Borror, Donald J., and Richard E. White. 1970. A Field Guide to the Insects of America North of Mexico. Boston: Houghton Mifflin Company.

Cobb, Boughton. 1963. A Field Guide to the Ferns. Boston, Houghton Mifflin Company.

Coon, Nelson. 1969. Using Wayside Plants. New York, Hearthside Press, Inc.

Courtney, Booth, and James Hall Zimmerman. 1972. Wildflowers and Weeds. New York: Van Nostrand Reinhold Company.

Crittenden, Mabel, and Dorothy Telfer. 1977. Wildflowers of the East. Millbrae, Calif.: Celestial Arts.

Cuthbert, Mabel Jaques. 1948. How to Know the Fall Flowers. Dubuque, Iowa: Wm. C. Brown Company.

Dana, Mrs. William Starr. 1963. How to Know the Wild Flowers. New York: Dover Publications, Inc.

Deam, C. C. 1940. Flora of Indiana. Indianapolis: Department of Conservation, Division of Forestry.

Durant, Mary. 1976. Who Named the Daisy? Who Named the Rose? A Roving Dictionary of North American Wild Flowers. New York, Congdon and Weed, Inc.

Fernald, M. L. 1950. Gray's Manual of Botany; 8th ed. New York, American Book Company.

Gleason, Henry A., and A. Cronquist. 1963. Manual of Vascular Plants of North-eastern United States and Adjacent Canada. Princeton, N.J.: D. Van Nostrand Company, Inc.

Heiser, C. B., Jr., and Jack Humbles. 1966. Higher Plants. *In* Natural Features of Indiana, ed. A. A. Lindsey. Indianapolis: Indiana Academy of Sciences, pp. 256–63.

Homoya, Michael A. et al. 1985. Natural Regions of Indiana. Indiana Academy of Sciences, 94:245–68.

Indiana Department of Natural Resources. 1985. Directory of Indiana's Dedicated Nature Preserves. Indianapolis.

Klots, Alexander B. 1951. A Field Guide to the Butterflies of North America, East of the Great Plains. Boston, Mass.: Houghton Mifflin Company.

Kluger, Marilyn. 1970. The Wild Flavor. New York: Coward, McCann and Geoghegan, Inc.

Lemmon, Robert S., and Charles C. Johnson. 1961. Wildflowers of North America in Full Color. Garden City, N.J.: Hanover House.

Le Strange, Richard. 1977. A History of Herbal Plants. New York: Arco Publishing Company, Inc.

Little, Elbert L. 1980. Audubon Society Field Guide to North American Trees, Eastern Region. New York: Alfred A. Knopf, Inc.

Lucas, Richard. 1972. The Magic of Herbs in Daily Living. West Nyack, New York: Parker Publishing Company.

Marcin, Marietta Marshall. 1983. The Complete Book of Herbal Teas. New York: Congdon and Weed.

Miller, Orson K., Jr. 1972. Mushrooms of North America. New York: E. P. Dutton and Company, Inc.

Newcomb, Lawrence. 1977. Newcomb's Wildflower Guide. Boston: Little, Brown and Company.

Niering, William A., and Nancy C. Olmstead. 1979. The Audubon Society Field Guide to North American Wildflowers, Eastern Region. New York: Alfred A. Knopf, Inc.

Peterson, Lee Allen. 1977. A Field Guide to Edible Wild Plants of Eastern and Central North America. Boston: Houghton Mifflin Company.

Peterson, Roger Tory, and Margaret McKenny. 1968. A Field Guide to Wildflowers of Northeastern and Northcentral North America. Boston: Houghton Mifflin Company.

Petty, Robert O, and Marion T. Jackson. 1966. Plant Communities. *In* Natural Features of Indiana, ed. A. A. Lindsey. Indianapolis: Indiana Academy of Sciences, pp. 264–96.

Price, Molly. 1966. The Iris Book. Princeton, N.J.: D. Van Nostrand Company, Inc.

Runkel, Sylvan T., and Alvin F. Bull. 1979. Wildflowers of Indiana Woodlands. Des Moines, Iowa: Wallace Homestead Book Company.

Smith, Alexander. 1958. The Mushroom Hunter's Field Guide. Ann Arbor: The University of Michigan Press.

Smith, Helen V. 1961. Michigan Wildflowers. Bloomfield Hills, Mich.: Cranbrook Institute of Science.

Stokes, Donald W., and Lillian Q. Stokes. 1985. A Guide to Enjoying Wildflowers. Boston: Little, Brown and Company.

Stupka, Arthur. 1965. Wildflowers in Color. New York: Harper and Row, Publishers.

Taylor, Kathryn S., and Stephen F. Hamblin. 1963. Handbook of Wild Flower Cultivation. New York: Macmillan Publishing Company.

Verhoek, Susan, and Mabel Jaques Cuthbert. 1982. How to Know the Spring Flowers. Dubuque, Iowa: Wm. C. Brown Company.

Wharton, Mary E., and Roger W. Barbour. 1971. A Guide to the Wildflowers and Ferns of Kentucky. Lexington: The University Press of Kentucky.

Williams, John G., and Andrew E. Williams. 1983. Field Guide to Orchids of North America from Alaska, Greenland, and the Arctic South to the Mexican Border. New York: Universe Books.

ACKNOWLEDGMENTS

Fred and Maryrose Wampler, together with Indiana University Press, want to acknowledge publicly the generous support that has made this book possible. Listed below are the names of individuals and companies who have either purchased paintings and permitted their reproduction in this volume or helped underwrite the costs of production by sponsoring various plates. It is sometimes said that art is not an essential commodity in our daily lives. We know otherwise. We want to thank the owners and sponsors who recognize this truth and helped make this beautiful book a reality.

OWNERS

Tom and Lela Asay, Nashville, TN 8
Mary Eleanor Cinkoske, Bloomington 9
Ruth M. Davison, Bloomington 1, 6, 20, 69, 80
Sandra J. DeWeese, Bloomington 16
Roberta L. Diehl, Bloomington 2
Sterling and Melinda Doster, Bloomington 4, 5
Jerry and Susan Gates, Bloomington 3, 50
Don and June Hendricks, Bloomington 19, 76
E. Robert Hunsicker, Bloomington 22
David Johnloz, Bloomington 13
Vernon and Diane Kliever, Bloomington 18, 29, 30
Emanuel and Judith Klein, Bloomington 43
David and Lisa Kratzer, Los Alamos, NM 17, 38
Bill and Janice Lacefield, Indianapolis 62
James and Kita Moore, Crawfordsville 57, 58
Holly and Ryan Mull, Bloomington 14
Loren and Gertrude Shahin, Columbus 44
Fred Voss, Westwood, NJ 11
Donald and Jacqueline Wade, Colorado Springs, CO 72
Martha Wade, San Francisco, CA 74
William R. Wampler, Bloomington 77
Ruth Wampler, Bloomington 70
James Way, M.D., Bloomington 25
Doug and Carol Weir, Mitchell 66
Marion and Mary Young, Bloomington 45

SPONSORS

Susan M. Anderson, M.D. and Neil E. Irick, M.D., Indianapolis 19
Robert D. Arnold, M.D., Indianapolis 3
Neva Y. Arnold, Indianapolis 37
Ruby Higginson Au, Prospect, KY 41
Mr. and Mrs. Robert P. Bareikis, Los Alamitos, CA 20
John and Peggy Bender, Bloomington 45
Jeffrey N. and Sarah A. Bush, Columbus 6
Kathleen Ketterman and Carol de Saint Victor, In Memory of Our Fathers, C. W. Ketterman and Douglas Hudson 57
Dr. and Mrs. G. Thomas Childes, Bloomington 14
John and Susan Cronkhite, Bloomington 65
Bart and Cinda Culver, North Webster 1
Charles and Harriet Curry, Bloomington 27
Dr. and Mrs. Gilbert S. Daniels, Indianapolis 67
Reed and Jane Dickerson, Bloomington 78
Michael J. and Willetta M. Ellis, Bloomington 26
Joan and James Ferguson, Bloomington 71
Stephen W. and Elaine Ewing Fess, Zionsville 56
In Memory of Samuel Parker Fields by Richard and Myrna Fields 75
Dr. and Mrs. David Frey, Bloomington 2
The Garden Club of Indiana, Inc. 59
Mr. and Mrs. W. W. Gasser, Jr., Portage 10
Mr. and Mrs. Raymond Gray, Nashville 35
George and Vera Greene, Bloomington 13
Harrell, Clendening & Coyne, Bloomington 50
Roger and Nancy Heimlich, Columbus 55
Mr. and Mrs. James R. Herdrich, Connersville 7
Indiana Heritage Arts, Inc., Nashville 63
Ruth and Dick Johnson, Columbus 53
Burton E. and Eleanor P. Jones, Bloomington 79
Laura Kern, Rochester 74
The Doctors Judith and Emanuel Klein, Bloomington 70
Kay F. Koch, North Salem 5
Janice F. and William B. Lacefield, Indianapolis 43

Janice F. and William B. Lacefield, Indianapolis 47
Janice F. and William B. Lacefield, Indianapolis 76
Mrs. Frank M. Lucas, Anderson 34
LTC and Mrs. H. M. Lynch, Bloomington 30
James and Amy Mason, West Terre Haute 12
Dr. and Mrs. Charles Matsumoto, Indianapolis 39
Marian Y. Meditch, Indianapolis 49
Metropolitan Printing Service, Inc., Bloomington 66
Charles I. and Ruth P. Miller, West Lafayette 4
Charles I. and Ruth P. Miller, West Lafayette 46
Nancy Ann Miller, Bloomington 62
James and Kita Moore, In Memory of Ivan T. Bauman (1914–1985) 58
Dr. and Mrs. Noel T. Moore, Hagerstown 15
Patricia Newforth, In Honor of Flona Query and Other Flowers 32
Judith G. Palmer, In Memory of My Grandparents, Mr. and Mrs. Donald B. Routt 72
Dr. and Mrs. Willis W. Peelle, Kokomo 11
Jill Perelman, Carmel 25
Mrs. Bernard Perry, Bloomington 18
Donna Polyak, Beverly Shores 42
Mrs. Millard F. Purcell, Shelbyville 52
Barbara Feucht Randall, Bloomington 69
Mildred A. Reeves, Columbus 48
Dr. and Mrs. Robert L. Reid, Evansville 29
Mr. and Mrs. Ralph D. Richards, Carmel 21
Mr. and Mrs. Ralph D. Richards, Carmel 16
Jane Rodman, Bloomington 77
Dr. and Mrs. Frank Shahbahrami, Bloomington 8
Mr. and Mrs. Loren E. Shahin, Columbus 44
Dr. Barbara Shalucha, Springfield, VT 28
Shand Mining, Inc., Carmel 9
George D. Smith, Jr., Indianapolis 7
R. Q. and Lou Thompson, Greenwood 80
Herman B Wells, Bloomington 23
Herman B Wells, Bloomington 51
Mrs. Milo Wells, Spencer 73
Olive Elizabeth White, Medaryville 22
Mr. and Mrs. Robert E. Williams, Charlestown 31
Donald R. and Frances D. Winslow, Bloomington 38

Matching gifts were provided by Eli Lilly and Company, Indianapolis.

INDEX

Other common names as well as the species name are provided for each entry. Main plate numbers for each species are in bold face.

Achillea millefolium, White Yarrow, **43**
Actaea pachypoda, White Baneberry, **15**
Actinomeris alternifolia, Wingstem, **63**
Adam and Eve, Puttyroot Orchid, *Aplectrum hyemale*, **26**
Adders-tongue, Trout Lily, Fawn Lily, Dog-tooth Violet, *Erythronium americanum*, **6**
Agastache scrophulariaefolia, Giant Hyssop, **72**
Aletris farinosa, Colicroot, Star Grass, **49**
Alexanders, Golden, *Zizia aurea*, **8**
Allium stellatum, Wild Onion, **47**
Allium vineale, Field Garlic, **47**
Althaea officinalis, Marsh-mallow, **40**
Alumroot, *Heuchera americana*, **18**
Alyssum, Hoary, *Berteroa incana*, **35**
American Cowslip, Shooting Star, Mosquito Bills, Birdbills, Prairie Pointers, *Dodecatheon meadia*, **11**
American Ipecac, *Gillenia stipulata*, **37**
American Lotus, *Nelumbo lutea*, **67**
Amorpha canescens, Leadwort, Prairie Shoestring, **52**
Anagallis arvensis, Scarlet Pimpernel, **44**
Anemone, False Rue, *Isopyrum biternatum*, **15**
Anemone quinquefolia, Wood Anemone, **22**
Anemone, Rue, *Anemonella thalictroides*, **4**
Anemone virginiana, Thimbleweed, **57**
Anemone, Wood, *Anemone quinquefolia*, **22**
Anemonella thalictroides, Rue Anemone, **4**
Antennaria plantaginifolia, Plantain-leaved Pussytoes or Early Ever-lastings, **11**
Apios americana, Groundnut, **70**
Aplectrum hyemale, Puttyroot Orchid, Adam and Eve, **26**
Apocynum androsaemifolium, Spreading Dogbane, **44**
Apocynum cannabinum, Indian Hemp, Hemp Dogbane, **62**
Appendaged Waterleaf, *Hydrophyllum appendiculatum*, **24**
Aquilegia canadensis, Wild Columbine, **10**
Arabis laevigata, Smooth Rock Cress, **10**
Arbutus, Trailing or Mayflower, *Epigaea repens*, **2**
Arisaema dracontium, Green Dragon Plant, **26**
Arisaema triphyllum, Jack-in-the-pulpit, **14;** green form, **15**
Asarum canadense, Wild Ginger, **11**
Asclepias amplexicaulis, Blunt-leaved Milkweed, **34**
Asclepias hirtella, Green Milkweed, **60**
Asclepias incarnata, Swamp Milkweed, **64**
Asclepias quadrifolia, Four-leaved Milkweed, **31**
Asclepias syriaca, Common Milkweed, **42**
Asclepias tuberosa, Butterfly Weed, **62**
Aster cordifolius, Blue Wood or Heart-leaved Aster, **79**
Aster pilosus, White Fall Aster, **80**
Aster puniceus, Purple-stemmed or Swamp Aster, **78**
Aster umbellatus, Flat-topped White Aster, **73**
Aster, Blue Wood or Heart-leaved, *Aster cordifolius*, **79**
Aster, Flat-topped White, *Aster umbellatus*, **73**
Aster, Heart-leaved or Blue Wood, *Aster cordifolius*, **79**
Aster, Maryland Golden, *Chrysopsis mariana*, **56**
Aster, Purple-stemmed or Swamp, *Aster puniceus*, **78**
Aster, White Fall, *Aster pilosus*, **80**
Aureolaria flava, False Foxglove, **71**
Avens, White, *Geum canadense*, **50**

Baneberry, White, *Actaea pachypoda*, **15**
Baptisia leucantha, Prairie or White False Indigo, **48**
Barbarea vulgaris, Winter Cress, Yellow Rocket, **8**
Basil Balm, White Monarda, *Monarda clinopodia*, **40**
Beard-tongue, *Penstemon calycosus*, **39**
Bedstraw, Northern, *Galium boreale*, **35**
Bee-balm, Oswego Tea, *Monarda didyma*, **46**
Beggar's-lice, Hound's tongue, *Cynoglossum officinale*, **36**
Bellflower, Tall, *Campanula americana*, **65**
Bellwort, *Uvularia perfoliata*, **15**
Bent or Drooping Trillium, *Trillium flexipes*, **13**
Bergamot, Wild, Monarda, *Monarda fistulosa*, **53**
Berteroa incana, Hoary Alyssum, **35**
Bidens aristosa, Tickseed Sunflower, **69**
Biennial Gaura, *Gaura biennis*, **71**
Bindweed, Hedge, *Convolvulus sepium*, **38, 70**
Bird's Nest, Queen Anne's Lace, Wild Carrot, *Daucus carota*, **58**
Birdbills, Shooting Star, Mosquito Bills, Prairie Pointers, American Cowslip, *Dodecatheon meadia*, **11**
Birdfoot Trefoil, *Lotus corniculatus*, **36**
Birdfoot Violet, *Viola pedata*, **12**
Bishop's Cap, Miterwort, *Mitella diphylla*, **21**
Bitter-bloom, Rose Pink, Rose Gentian, *Sabatia angularis*, **60**
Bittersweet, *Celastrus scandens*, **26,** 80
Blackberry, *Rubus* species, **27, 58**
Black-eyed Susan, *Rudbeckia hirta*, **62**
Black Raspberry, Thimbleberry, *Rubus* species, **43**
Blazing Star, Gayfeather, *Liatris scariosa*, **73**
Blazing Star, Dense, Gayfeather, *Liatris spicata*, **73**
Blazing Star, Rough, *Liatris aspera*, **73**
Bleeding Heart, 6
Blephilia ciliata, Downy Wood-mint, **39**
Bloodroot, *Sanguinaria canadensis*, **5**
Blue Cohosh, *Caulophyllum thalictroides*, **14**
Blue-eyed Grass: *Sisyrinchium angustifolium*, **42** *Sisyrinchium montanum*, **12**
Blue-eyed Mary, *Collinsia verna*, **7**
Blue Flag, *Iris versicolor*, **38**
Blue Lettuce, *Lactuca floridana*, **70**
Blue or Great Lobelia, *Lobelia siphilitica*, **78**
Blue Mistflower, *Eupatorium coelestinum*, **71**
Blue Phlox, Wild Sweet William, *Phlox divaricata*, **16**
Blue Vervain, *Verbena hastata*, **53**
Blue or Cow Vetch, *Vicia cracca*, **41**
Blue Violet, Common, Meadow Violet, *Viola papilionacea*, **14, 16**

Blue Wood or Heart-leaved Aster, *Aster cordifolius*, **79**
Bluebells, Virginia, Virginia Cowslip, *Mertensia virginica*, **16**
Bluet, Summer, Large Houstonia, *Houstonia purpurea*, **39**
Bluets, Innocence, Quaker Ladies, *Houstonia caerulea*, **11**
Blunt-leaved Milkweed, *Asclepias amplexicaulis*, **34**
Boneset, Common Throughwort, *Eupatorium perfoliatum*, **78**
Bottle or Closed Gentian, *Gentiana andrewsii*, **77**
Bouncing Bet, Soapwort, *Saponaria officinalis*, **65**
Broad-leaved Waterleaf, *Hydrophyllum canadense*, **40**
Brook Lobelia, *Lobelia kalmii*, **76**
Brown County State Park, 42
Bush Clover, Roundheaded, *Lespedeza capitata*, **71**
Bush Clover, Wandlike, *Lespedeza intermedia*, **72**
Butter-and-eggs, Toadflax, *Linaria vulgaris*, **46**
Buttercup, *Ranunculus hispidus*, **15**
Buttercup, Swamp, *Ranunculus septentrionalis*, **19**
Butterfly Weed, *Asclepias tuberosa*, **62**
Button Bush, *Cephalanthus occidentalis*, **47**
Button Snakeroot, Rattlesnake-Master, *Eryngium yuccifolium*, **48**

Caltha palustris, Marsh-marigold, Cowslip, **21**
Camassia scilloides, Wild Hyacinth, **23**
Campanula americana, Tall Bellflower, **65**
Campion, Starry, *Silene stellata*, **53**
Campsis radicans, Trumpet Vine, **57**
Canada Mayflower, Wild Lily of the Valley, *Maianthemum canadense*, **32**
Canadian Thistle, *Cirsium arvense*, **49**
Cardamine bulbosa, Spring Cress, **21**
Cardamine douglassii, Purple Cress, **9**
Cardinal Flower, *Lobelia cardinalis*, **69**
Carduus nutans, Nodding or Musk Thistle, **41**
Carrot, Wild, Queen Anne's Lace, Bird's Nest, *Daucus carota*, **58**
Cassia fasciculata, Partridge Pea, **61**
Cassia hebecarpa, Wild Senna, **63**
Catchfly, Fire Pink, *Silene virginica*, **28**
Catchfly, Night-Flowering, *Silene noctiflora*, **36**
Catchfly, Royal, *Silene regia*, **59**
Caulophyllum thalictroides, Blue Cohosh, **14**
Cedar Bluffs nature preserve, 3
Celandine, *Chelidonium majus*, **7**
Celandine Poppy, Wood Poppy, *Stylophorum diphyllum*, **7**
Celastrus scandens, Bittersweet, **26, 80**
Centaurea maculosa, Spotted Knapweed, **44**
Cephalanthus occidentalis, Button Bush, **47**
Chain O'Lakes State Park, 51
Cheeses, *Malva neglecta*, **44**
Chelidonium majus, Celandine, **7**
Chelone glabra, Turtlehead, **77**
Chickweed, Star, *Stellaria pubera*, **16**
Chicory, *Cichorium intybus*, **58,** 65
Chrysanthemum leucanthemum, Ox-eye Daisy, **43**
Chrysopis mariana, Maryland Golden Aster, **56**
Cicely, Sweet, *Osmorhiza claytoni*, **22**
Cichorium intybus, Chicory, **58,** 65
Cinquefoil, Rough-fruited, Five Fingers, *Potentilla recta*, **55**
Cinquefoil, Trailing or Common, Five Fingers, *Potentilla simplex*, **28**
Cirsium altissimum, Tall Thistle, **79**
Cirsium arvense, Canadian Thistle, **49**
Claytonia virginica, Spring Beauty, 4, **6**
Cleavers, Goosegrass, *Galium aparine*, **22**
Cleft Phlox, Creeping or Sand Phlox, *Phlox bifida*, **9**
Clematis viorna, Leather Flower, **37**
Clifty Falls State Park, 5
Closed or Bottle Gentian, *Gentiana andrewsii*, **77**
Clover, Red, *Trifolium pratense*, **65**
Clover, Roundheaded Bush, *Lespedeza capitata*, **71**
Clover, Wand-like Bush, *Lespedeza intermedia*, **72**
Clover, White, *Trifolium repens*, **33**
Clover, Yellow Sweet, *Melilotus officinalis*, **46**
Coffee, Wild, Horse Gentian, *Triosteum aurantiacum*, **26**
Cohosh, Blue, *Caulophyllum thalictroides*, **14**
Colicroot, Star Grass, *Aletris farinosa*, **49**
Collinsia verna, Blue-eyed Mary, **7**
Columbine, Wild, *Aquilegia canadensis*, **10**
Commelina communis, Dayflower, **65**
Common Blue Violet, Meadow Violet, *Viola papilionacea*, **14, 16**
Common Daylily, *Hemerocallis fulva*, **46**
Common Evening Primrose, *Oenothera biennis*, **72**
Common Milkweed, *Asclepias syriaca*, **42**
Common Mullein, *Verbascum thapsus*, **57**
Common or Trailing Cinquefoil, Five Fingers, *Potentilla simplex*, **28**
Common Throughwort, Boneset, *Eupatorium perfoliatum*, **78**
Coneflower, Gray or Prairie, *Ratibida pinnata*, **54**
Confederate Violet, *Viola papilionacea var. priceana*, **9**
Convolvulus sepium, Hedge Bindweed, **38, 70**
Coreopsis tripteris, Tall Coreopsis, **72**
Coreopsis, Tall, *Coreopsis tripteris*, **72**
Corn Speedwell, *Veronica arvensis*, **10**
Coronilla varia, Crownvetch, **41**
Corpse Plant, Indian Pipes, *Monotropa uniflora*, **66**
Corydalis aurea, Golden Corydalis, **8**
Corydalis, Golden, *Corydalis aurea*, **8**
Cow or Blue Vetch, *Vicia cracca*, **41**
Cowslip, Marsh-marigold, *Caltha palustris*, **21**
Cowslip, Virginia, Virginia Bluebells, *Mertensia virginica*, **16**
Cranberry, Large, *Vaccinium macrocarpon*, **33**
Cranesbill, Wild Geranium, *Geranium maculatum*, **16**
Creeping Phlox, Sand or Cleft Phlox, *Phlox bifida*, **9**
Creeping Wood Sorrel, *Oxalis stricta*, **33**
Cress, Spring, *Cardamine bulbosa*, **21**
Cress, Winter, Yellow Rocket, *Barbarea vulgaris*, **8**
Cress, Purple, *Cardamine douglassii*, **9**
Crested Dwarf Iris, *Iris cristata*, **17**
Crimson-eyed Rose-mallow, *Hibiscus palustris, forma peckii*, **51,** 64
Crownbeard, Sunflower, *Verbesina helianthoides*, **37**
Crownvetch, *Coronilla varia*, **41**
Cucumber Root, Indian, *Medeola virginiana*, **32**
Culver's Root, *Veronicastrum virginicum*, **59**
Cut-leaved toothwort, *Dentaria laciniata*, **5**
Cynoglossum officinale, Hound's-tongue, Beggar's-lice, **36**
Cynthia, Two-flowered, Dwarf Dandelion, *Krigia biflora*, **18**
Cypripedium acaule, Pink Lady's-slipper, Moccasin Flower, 25, **30**
Cypripedium calceolus, Yellow Lady's-slipper: *var. pubescens*, **25;** *var. parviflorum*, **29**
Cypripedium candidum, White Lady's-slipper, **29**
Cypripedium reginae, Showy Lady's-slipper, **38**

Daisy Fleabane, *Erigeron annuus*, **43**
Daisy, Ox-eye, *Chrysanthemum leucanthemum*, **43**
Dame's Rocket, Sweet Rocket, Violet Rocket, *Hesperis matronalis*, **27,** 28
Dandelion, Dwarf, Two-flowered Cynthia, *Krigia biflora*, **18**
Dandelion, *Taraxacum officinale*, **8**
Datura stramonium, Jimsonweed or Thorn Apple, **79**
Daucus carota, Queen Anne's Lace, Wild Carrot, Bird's Nest, **58**
Dayflower, *Commelina communis*, **65**
Daylily, Common, *Hemerocallis fulva*, **46**

Deergrass, Meadow Beauty, *Rhexia virginica,* **68**
Delphinium tricorne, Dwarf or Spring Larkspur, **13, 14**
Dense Blazing Star, Gayfeather, *Liatris spicata,* **73**
Dentaria laciniata, Cut-leaved Toothwort, **5**
Deptford Pink, *Dianthus armeria,* **42**
Desmodium glutinosum, Pointed-leaved Tick Trefoil, **52**
Devil's Paintbrush, Orange Hawkweed, *Hieracium aurantiacum,* **36**
Devil's Shoestrings, Goat's Rue, Wild Sweet Pea, *Tephrosia virginiana,* **34**
Dianthus armeria, Deptford Pink, **42**
Dicentra canadensis, Squirrel Corn, **6**
Dicentra cucullaria, Dutchman's Breeches, **6**
Dipsacus sylvestris, Teasel, **44**
Dock, Prairie, *Silphium terebinthinaceum,* **54**
Dockmackie, Maple-leaved Viburnum, *Viburnum acerifolium,* **28**
Doctrine of Signatures, 4, 78
Dodecatheon meadia, Shooting Star, Mosquito Bills, Birdbills, Prairie Pointers, American Cowslip, **11**
Dog Violet, *Viola conspersa,* **21**
Dog-tooth Violet, Trout Lily, Fawn Lily, Adders-tongue, *Erythronium americanum,* **6**
Dogbane, Hemp, Indian Hemp, *Aposynum cannabinum,* **62**
Dogbane, Spreading, *Apocynum androsaemifolium,* **44**
Downy or Woolly Blue Violet, *Viola sororia,* **6**
Downy Rattlesnake Plantain, *Goodyera pubescens,* **74**
Downy Skullcap, *Scutellaria incana,* **50**
Downy Wood-mint, *Blephilia ciliata,* **39**
Downy Yellow Violet, *Viola pubescens,* **20**
Dragonhead, False, Obedient-plant, *Physostegia virginiana,* **73**
Drooping or Bent Trillium, *Trillium flexipes,* **13**
Dutchman's Breeches, *Dicentra cucullaria,* **6**, 8
Dwarf Dandelion, Two-flowered Cynthia, *Krigia biflora,* **18**
Dwarf Ginseng, *Panax trifolium,* **17**
Dwarf or Spring Larkspur, *Delphinium tricorne,* **13, 14**
Dwarf White Trillium, Snow Trillium, *Trillium nivale,* **3**

Early Everlastings, Plantain-leaved Pussytoes, *Antennaria plantaginifolia,* **11**
Early Goldenrod, *Solidago juncea,* **61**
Eastern Larch, Tamarack, *Larix laricina,* 30
Ebony Spleenwort Fern, 50
Elderberry, *Sambucus canadensis,* **79**
Epigaea repens, Trailing Arbutus, Mayflower, **2**
Erigenia bulbosa, Harbinger of Spring, Salt-and-pepper, **3**
Erigeron annuus, Daisy Fleabane, **43**
Eryngium yuccifolium, Rattlesnake-Master, Button Snakeroot, **48**
Erysimum cheiranthoides, Wormseed Mustard, **19**
Erythronium albidum, White Trout Lily, **6**
Erythronium americanum, Trout Lily, Fawn Lily, Adders-tongue, Dog-tooth Violet, **6**
Eupatorium coelestinum, Mistflower, Hardy Ageratum, **71**
Eupatorium fistulosum, Hollow Joe-Pye-Weed, **70**
Eupatorium maculatum, Spotted Joe-Pye-Weed, **76**
Eupatorium perfoliatum, Boneset, Common Throughwort, **78**
Eupatorium rugosum, White Snakeroot, White Mistflower, **68**
Euphorbia corollata, Flowering Spurge, **49**
Evening Lychnis, *Lychnis alba,* **80**
Evening Primrose, Common, *Oenothera biennis,* **72**
Everlasting Pea, Perennial Sweet Pea, *Lathyrus latifolius,* **41**

False Dragonhead, Obedient-plant, *Physostegia virginiana,* **73**
False Foxglove, *Aureolaria flava,* **71**
False Indigo, Prairie or White, *Baptisia leucantha,* **48**
False Rue Anemone, *Isopyrum biternatum,* **15**
False Solomon's Seal, *Smilacina racemosa,* 21, **24**
Fawn Lily, Trout Lily, Adders-tongue, Dog-tooth Violet, *Erythronium americanum,* **6**
Featherbells, Featherfleece, *Stenanthium gramineum,* **40**
Fern, Ebony Spleenwort, **50**
Field Garlic, *Allium vineale,* **47**
Field Hawkweed, King Devil, *Hieracium pratense,* **36**
Field Milkwort, *Polygala sanguinea,* **60**
Field Pansy, *Viola Kitaibeliana,* **10**
Fire Pink, Catchfly, *Silene virginica,* **28**
Five Fingers: Rough-fruited Cinquefoil, *Potentilla recta,* **55;** Trailing or Common Cinquefoil, *Potentilla simplex,* **28**
Flag, Blue, *Iris versicolor,* **38**
Flat-topped White Aster, *Aster umbellatus,* **73**
Fleabane, Daisy, *Erigeron annuus,* **43**
Flowering Spurge, *Euphorbia corollata,* **49**
Fogfruit, *Lippia lanceolata,* **47**
Forget-me-not, *Myosotis scorpioides,* **39**
Four-o'clock, Wild, Heart-leaved Umbrellawort, *Mirabilis nyctaginea,* **34**
Four-leaved Milkweed, *Asclepias quadrifolia,* **31**
Foxglove, False, *Aureolaria flava,* **71**
Fragaria virginiana, Wild Strawberry, **19**
Fragile Fern, 5
Fringed Gentian, *Gentiana crinita,* **76**

Galium aparine, Cleavers, Goosegrass, **22**
Galium boreale, Northern Bedstraw, **35**
Garlic, Field, *Allium vineale,* **47**
Gaura, Biennial, *Gaura biennis,* **71**
Gaura biennis, Biennial Gaura, **71**
Gayfeathers, Dense Blazing Star, *Liatris spicata,* **73**
Gentian, Closed or Bottle, *Gentiana andrewsii,* **77**
Gentian, Fringed, *Gentiana crinita,* **76**
Gentian, Horse, Wild Coffee, *Triosteum aurantiacum,* **26**
Gentian, Rose, Rose Pink, Bitter-bloom, *Sabatia angularis,* **60**
Gentiana andrewsii, Closed or Bottle Gentian, **77**
Gentiana crinita, Fringed Gentian, **76**
Geranium maculatum, Wild Geranium, Cranesbill, **16**
Geranium, Wild, *Geranium maculatum,* **16**
Gerardia, Purple, *Gerardia purpurea,* **76**
Gerardia purpurea, Purple Gerardia, **76**
German Iris, *Iris germanica,* **27**
Germander, Wood-sage, *Teucrium canadense,* **52**
Geum canadense, White Avens, **50**
Giant Hyssop, *Agastache scrophulariaefolia,* **72**
Gill-over-the-ground, Ground Ivy, *Glechoma hederacea,* **8**
Gillenia stipulata, American Ipecac, **37**
Ginger, Wild, *Asarum canadense,* **11**
Ginseng, Dwarf, *Panax trifolium,* **17**
Ginseng, *Panax quinquefolium,* **32**
Glechoma hederacea, Gill-over-the-ground, Ground Ivy, **8**
Goat's Rue, Wild Sweet Pea, Devil's Shoestrings, *Tephrosia virginiana,* **34**
Goatsbeard, Meadow Salsify, *Tragopogon pratensis,* **55**
Golden Alexanders, *Zizia aurea,* **8**
Golden Corydalis, *Corydalis aurea,* **8**
Golden Ragwort, Groundsel, Squaw-weed, *Senecio aureus,* 12, **21**
Goldenrod, Early, *Solidago juncea,* **61**
Goldenrod, Late, *Solidago gigantea,* **80**
Goldenrod, *Solidago species,* **75**
Goldenseal, Yellowroot, *Hydrastis canadensis:* flower, **20;** fruit, **50**
Goodyera pubescens, Downy Rattlesnake Plantain, **74**
Goosegrass, Cleavers, *Galium aparine,* **22**

Gray or Prairie Coneflower, *Ratibida pinnata,* **54**
Great or Blue Lobelia, *Lobelia siphilitica,* **78**
Greek Valerian, *Polemonium reptans,* **10**
Green Dragon Plant, *Arisaema dracontium,* **26**
Green Milkweed, *Asclepias hirtella,* **60**
Green Violet, *Hybanthus concolor,* **22**
Ground-cedar, *Lycopodium flabelliforme,* **74**
Ground Cherry, *Physalis heterophylla,* **55**
Ground Ivy, Gill-over-the-ground, *Glechoma hederacea,* **8**
Groundnut, *Apios americana,* **70**
Groundsel, Golden Ragwort, Squaw-weed, *Senecio aureus,* 12, **21**

Habenaria peramoena, Purple Fringeless Orchid, **60**
Hairy Hawkweed, *Hieracium gronovii,* **49**
Hairy Puccoon, *Lithospermum croceum,* **34**
Harbinger of Spring, Salt-and-pepper, *Erigenia bulbosa,* **3**
Hardhack, Steeplebush, *Spiraea tomentosa,* **53**
Hardy Ageratum, Mistflower, *Eupatorium coelestinum,* **71**
Harmonie State Park 10, 56
Hawkweed, Field, King Devil, *Hieracium pratense,* **36**
Hawkweed, Hairy, *Hieracium gronovii,* **49**
Hawkweed, Orange, Devil's Paintbrush, *Hieracium aurantiacum,* **36**
Hawkweed, Panicled, *Hieracium paniculatum,* **31**
Heal-all or Selfheal, *Prunella vulgaris,* **47**
Heart-leaved or Blue Wood Aster, *Aster cordifolius,* **79**
Heart-leaved Umbrellawort, Wild Four-o'clock, *Mirabilis nyctaginea,* **34**
Hedge Bindweed, *Convolvulus sepium,* **38, 70**
Helenium autumnale, Sneezeweed, **75**
Helenium nudiflorum, Purple-headed Sneezeweed, **61,** 75
Helianthus divaricatus, Woodland Sunflower, **63**
Helianthus tuberosus, Jerusalem Artichoke, **80**
Hemerocallis fulva, Common Daylily, **46**
Hemp Dogbane, Indian Hemp, *Apocynum cannabinum,* **62**
Henbit, *Lamium amplexicaule,* **10**
Hepatica acutiloba, Hepatica, **4**
Hepatica americana, 4
Hepatica, *Hepatica acutiloba,* **4**
Hesperis matronalis, Dame's Rocket, Sweet Rocket, Violet Rocket, **27,** 28
Heuchera americana, Alumroot, **18**
Hibiscus palustris, Swamp Rose-mallow, **64;** *forma peckii,* Crimson-eyed Rose-mallow, **51,** 64
Hieracium aurantiacum, Orange Hawkweed, Devil's Paintbrush, **36**
Hieracium gronovii, Hairy Hawkweed, **49**
Hieracium paniculatum, Panicled Hawkweed, **31**
Hieracium pratense, Field Hawkweed, King Devil, 36
Hoary Alyssum, *Berteroa incana,* **35**
Hoary Puccoon, *Lithospermum canescens,* **12**
Hoary Vervain, *Verbena stricta,* **58**
Hollow Joe-Pye-Weed, *Eupatorium fistulosum,* **70**
Honeysuckle, Japanese, *Lonicera japonica,* **46**
Horse Gentian or Wild Coffee, *Triosteum aurantiacum,* **26**
Horsemint, *Monarda punctata,* **52**
Hound's-tongue, Beggar's-lice, *Cynoglossum officinale,* **36**
Houstonia caerulea, Bluets, Innocence, Quaker Ladies, **11**
Houstonia purpurea, Summer Bluet, Large Houstonia, **39**
Houstonia, Large, Summer Bluet, *Houstonia purpurea,* **39**
Hovey Lake, 68
Hyacinth, Wild, *Camassia scilloides,* **23**
Hybanthus concolor, Green Violet, **22**
Hybrid Loosestrife, *Lysimachia hybrida,* **62**
Hydrastis canadensis, Goldenseal, Yellowroot: flower, **20;** fruit, **50**
Hydrophyllum appendiculatum, Appendaged Waterleaf, **24**
Hydrophyllum canadense, Broad-leaved Waterleaf, **40**
Hydrophyllum virginianum, Virginia Waterleaf, **13**
Hypericum perforatum, St. John's-wort, **57**
Hypoxis hirsuta, Stargrass, **12**
Hyssop, Giant, *Agastache scrophulariaefolia,* **72**

Impatiens capensis, Spotted Jewelweed, Touch-me-not, **63**
Impatiens pallida, Pale Jewelweed, Pale Touch-me-not, 63, **78**
Indian Cucumber Root, *Medeola virginiana,* **32**
Indian Hemp or Hemp Dogbane, *Apocynum cannabinum,* **62**
Indian Pink, Pink-root, Star-bloom, Worm-grass, *Spigelia marilandica,* **37**
Indian Pipes, Corpse Plant, *Monotropa uniflora,* **66**
Indian Tobacco, *Lobelia inflata,* **68**
Indiana Dunes State Park, 30, 48, 65
Indigo, Prairie or White False, *Baptisia leucantha,* **48**
Indigo, White or Prairie False, *Baptisia leucantha,* **48**
Innocence, Bluets, Quaker Ladies, *Houstonia caerulea,* **11**
Iodanthus pinnatifidus, Purple Rocket, **28**
Ipecac, American, *Gillenia stipulata,* **37**
Ipomoea hederacea, Ivy-leaved Morning-glory, **80**
Ipomoea pandurata, Wild Potato-vine, Purple-throated Morning Glory, **56**
Iris brevicaulis, Short-stemmed Iris, **31**
Iris cristata, Crested Dwarf Iris, **17**
Iris versicolor, Blue Flag, **38**
Iris, Crested Dwarf, *Iris cristata,* **17**
Iris, German, *Iris germanica,* **27**
Iris, Short-stemmed, *Iris brevicaulis,* **31**
Ironweed, Missouri, *Vernonia missurica,* **75**
Ironweed, Tall, *Vernonia altissima,* **79**
Isopyrum biternatum, False Rue Anemone, **15**
Ivy-leaved Morning Glory, *Ipomoea hederacea,* **80**

Jack-in-the-pulpit, *Arisaema triphyllum,* **14**; green form, **15**
Japanese Honeysuckle, *Lonicera japonica,* **46**
Jeffersonia diphylla, Twinleaf, **5**
Jeruslem Artichoke, *Helianthus tuberosus,* **80**
Jewelweed, Pale, Pale Touch-me-not, *Impatiens pallida,* 63, **78**
Jewelweed, Spotted, Touch-me-not, *Impatiens capensis,* **63**
Jimsonweed, Thorn Apple, *Datura stramonium,* **79**
Joe-Pye-Weed, Hollow, *Eupatorium fistulosum,* **70**
Joe-Pye-Weed, Spotted, *Eupatorium maculatum,* **76**

King Devil, Field Hawkweed, *Hieracium pratense,* **36**
Knapweed, Spotted, *Centaurea maculosa,* **44**
Krigia biflora, Two-flowered Cynthia, Dwarf Dandelion, **18**

Lactuca canadensis, Yellow Wild Lettuce, **46**
Lactuca floridana, Blue Lettuce, **70**
Lady's-slipper, Pink, Moccasin Flower, *Cypripedium acaule,* 25, **30**
Lady's-slipper, Showy, *Cypripedium reginae,* **38**
Lady's-slipper, White, *Cypripedium candidum,* **29**
Lady's-slipper, Yellow, *Cypridium calceolus, var. parviflorum,* **29;** *var. pubescens,* **25**
Lady's Thumb, Redleg, Smartweed, *Polygonum persicaria,* **70**
Lady's Tresses, Spiral Orchid, *Spiranthes vernalis,* **47**
Lady's Tresses, Little, *Spiranthes tuberosa,* **74**
Lady's Tresses, Nodding, *Spiranthes cernua,* **76**

Lamium amplexicaule, Henbit, **10**
Lance-leaved Loosestrife, *Lysimachia lanceolata*, **45**
Lance-leaved or Water Violet, *Viola lanceolata*, **33**
Large Cranberry, *Vaccinium macrocarpon*, **33**
Large Houstonia, Summer Bluet, *Houstonia purpurea*, **39**
Large Twayblade, Lily-leaved Twayblade, *Liparis lilifolia*, **18,** 25
Larix laricina, Tamarack, Eastern Larch, 30
Larkspur, Dwarf or Spring, *Delphinium tricorne*, **13, 14**
Late Goldenrod, *Solidago gigantea*, **80**
Lathyrus latifolius, Perennial Sweet Pea, Everlasting Pea, **41**
Leadwort, Prairie Shoestring, *Amorpha canescens*, **52**
Leather Flower, *Clematis viorna*, **37**
Liparis lilifolia, Large Twayblade, Lily-leaved Twayblade, **18,** 25
Lespedeza capitata, Roundheaded Bush Clover, **71**
Lespedeza intermedia, Wand-like Bush Clover, **72**
Lettuce, Blue, *Latuca floridana*, **70**
Lettuce, Yellow Wild, *Lactuca canadensis*, **46**
Liatris aspera, Rough Blazing Star, **73**
Liatris scariosa, Blazing Star, Gayfeather, **73**
Liatris spicata, Dense Blazing Star, Gayfeather, **73**
Lilium michiganense, Michigan Lily, **48**
Lilium superbum, Turk's cap, 48
Lily, Michigan, *Lilium michiganense*, **48**
Lily-leaved Twayblade, Large Twayblade, *Liparis lilifolia*, **18,** 25
Lily of the Valley, Wild, Canada Mayflower, *Maianthemum canadense*, **32**
Linaria vulgaris, Butter-and-eggs, Toadflax, **46**
Lincoln State Park, 58, 78
Lippia lanceolata, Fogfruit, **47**
Lithospermum canescens, Hoary Puccoon, **12**
Lithospermum croceum, Hairy Puccoon, **34**
Little Lady's Tresses, *Spiranthes tuberosa*, **74**
Liver-leaf (*Hepatica* sp.), 4
Lizard's Tail, *Saururus cernuus*, **51**
Lobelia cardinalis, Cardinal Flower, **69**
Lobelia inflata, Indian Tobacco, **68**
Lobelia kalmii, Brook Lobelia, **76**
Lobelia siphilitica, Blue or Great Lobelia, **78**
Lobelia spicata, Pale Spiked Lobelia, **49**
Lobelia, Blue or Great, *Lobelia siphilitica*, **78**
Lobelia, Brook, *Lobelia kalmii*, **76**
Lobelia, Pale Spiked, *Lobelia spicata*, **49**
Lonicera japonica, Japanese Honeysuckle, **46**
Loosestrife, Hybrid, *Lysimachia hybrida*, **62**
Loosestrife, Lance-leaved, *Lysimachia lanceolata*, **45,** 62
Loosestrife, Purple, Lythrum, *Lythrum salicaria*, **67**
Lotus, American, *Nelumbo lutea*, **67**
Lotus corniculatus, Birdfoot Trefoil, **36**
Lousewort, Wood-betony, *Pedicularis canadensis*, **20**
Ludwigia alternifolia, Seedbox, **61**
Lupine, Wild, *Lupinus perennis*, **35**
Lupinus perennis, Wild Lupine, **35**
Lychnis alba, Evening Lychnis, **80**
Lychnis, Evening, *Lychnis alba*, **80**
Lycopodium flabelliforme, Ground-cedar, **74**
Lysimachia hybrida, Hybrid Loosestrife, **62**
Lysimachia lanceolata, Lance-leaved Loosestrife, **45**
Lythrum salicaria, Lythrum, Purple Loosestrife, **67**

Maianthemum canadense, Canada Mayflower, Wild Lily of the Valley, **32**
Malva neglecta, Cheeses, **44**
Mandrake or Mayapple, *Podophyllum peltatum*, **20**
Maple-leaved Viburnum, Dockmackie, *Viburnum acerifolium*, **28**
Marsh-mallow, *Althaea officinalis*, **40**
Marsh-marigold, Cowslip, *Caltha palustris*, **21**
Martin State Forest, 50
Maryland Golden Aster, *Chrysopsis mariana*, **56**
Mayapple, Mandrake, *Podophyllum peltatum*, **20**
Mayflower, Canada, Wild Lily of the Valley, *Maianthemum canadense*, **32**
Mayflower, Trailing Arbutus, *Epigaea repens*, **2**
McCormick's Creek State Park, 11, 13, 15, 24, 66
Meadow Beauty, Deergrass, *Rhexia virginica*, **68**
Meadow Rue, *Thalictrum dioicum*, **11,** 14
Meadow Salsify, Goatsbeard, *Tragopogon pratensis*, **55**
Meadow Violet, Common Blue Violet, *Viola papilionacea*, **14, 16**
Meadowsweet, *Spiraea latifolia*, **52**
Medeola virginiana, Indian Cucumber Root, **32**
Melilotus officinalis, Yellow Sweet Clover, **46**
Mertensia virginica, Virginia Bluebells, Virginia Cowslip, **16**
Miami Mist, *Phacelia purshii*, **23**
Michigan Lily, *Lilium michiganense*, **48**
Milkweed, Blunt-leaved, *Asclepias amplexicaulis*, **34**
Milkweed, Common, *Asclepias syriaca*, **42**
Milkweed, Four-leaved, *Asclepias quadrifolia*, **31**
Milkweed, Green, *Asclepias hirtella*, **60**
Milkweed, Swamp, *Asclepias incarnata*, **64**
Milkwort, Field, *Polygala sanguinea*, **60**
Mimulus alatus, Sharp-winged Monkey-flower, **68**
Mimulus ringens, Monkey-flower, **53,** 68
Mint, Narrow Leaved Mountain, *Pycnanthemum tenuifolium*, **49**
Mirabilis nyctaginea, Wild Four-o'clock, Heart-leaved Umbrellawort, **34**
Missouri Ironweed, *Vernonia missurica*, **75**
Missouri Violet, *Viola missouriensis*, **7**
Mistflower, White, White Snakeroot, *Eupatorium rugosum*, **68**
Mistflower, Blue, Hardy Ageratum, *Eupatorium coelestinum*, 68, **71**
Mitella diphylla, Miterwort, Bishop's Cap, **21**
Miterwort, Bishop's Cap, *Mitella diphylla*, **21**
Moccasin Flower, Pink Lady's-slipper, *Cypripedium acaule*, 25, **30**
Monarda clinopodia, White Monarda, Basil Balm, **40**
Monarda didyma, Bee-balm, Oswego Tea, **46**
Monarda fistulosa, Monarda, Wild Bergamot, **53**
Monarda punctata, Horsemint, **52**
Monarda, White, Basil Balm, *Monarda clinopodia*, **40**
Monarda, Wild Bergamot, *Monarda fistulosa*, **53**
Monkey-flower, Sharp-winged, *Mimulus alatus*, **68**
Monkey-flower, *Mimulus ringens*, **53**
Monotropa uniflora, Indian Pipes, Corpse Plant, **66**
Morchella esculenta, Morel, 6
Morchella semilibera, Morel, 13
Morel, *Morchella esculenta*, 6
Morel, *Morchella semilibera*, 13
Morning Glory, Ivy-leaved, *Ipomoea hederacea*, **80**
Morning Glory, Purple-throated, Wild Potato Vine, *Ipomoea pandurata*, **56**
Mosquito Bills, Shooting Star, Birdbills, Prairie Pointers, American Cowslip, *Dodecatheon meadia*, **11**
Moth Mullein, *Verbascum blattaria*, **55**
Mounds State Park, 26
Mountain Mint, Narrow Leaved, *Pycnanthemum tenuifolium*, **49**
Mullein, Common, *Verbascum thapsus*, **57**
Mullein, Moth, *Verbascum blattaria*, **55**
Multiflora Rose, *Rosa multiflora*, **27**
Musk Thistle, Nodding Thistle, *Carduus nutans*, **41**
Mustard, Wormseed, *Erysimum cheiranthoides*, **19**
Myosotis scorpioides, Forget-me-not, **39**
Myrtle, Periwinkle, *Vinca minor*, **8**

Narrow Leaved Mountain mint, *Pycnanthemum tenuifolium*, **49**
Nelumbo lutea, American Lotus, **67**

Night-Flowering Catchfly, *Silene noctiflora,* **36**
Nodding Lady's Tresses, *Spiranthes cernua,* **76**
Nodding Pogonia, Three-birds Orchid, *Triphora trianthophora,* **66**
Nodding Thistle, Musk Thistle, *Carduus nutans,* **41**
Nodding Trillium, *Trillium cernuum,* 13
Northern Bedstraw, *Galium boreale,* **35**
Northern Downy Violet, *Viola fimbriatula,* **19**
Northern Pitcher Plant, *Sarracenia purpurea,* **30**
Nymphaea odorata, White Water Lily, **51**

Obedient-plant, False Dragonhead, *Physostegia virginiana,* **73**
Obolaria virginica, Pennywort, **12**
Oenothera biennis, Common Evening Primrose, **72**
Oenothera fruticosa, Sundrops, **45**
Onion, Wild, *Allium stellatum,* **47**
Orange Hawkweed, Devil's Paintbrush, *Hieracium aurantiacum,* **36**
Orchid, Purple Fringeless, *Platanthera peramoena, Habenaria peramoena,* **60**
Orchid, Puttyroot, Adam and Eve, *Aplectrum hyemale,* **26**
Orchid, Spiral, *Spiranthes vernalis,* **47**
Orchid, Three-birds, Nodding Pogonia, *Triphora trianthophora,* **66**
Orchis, Showy, *Orchis spectabilis,* **25**
Orchis spectabilis, Showy Orchis, **25**
Ornithogalum umbellatum, Star of Bethlehem, **8**
Osmorhiza claytoni, Sweet Cicely, **22**
Oswego Tea, Bee-balm, *Monarda didyma,* **46**
Ouabache State Park, **57**
Ox-eye Daisy, *Chrysanthemum leucanthemum,* **43**
Oxalis grandis, Yellow Oxalis, Wood Sorrel, **28**
Oxalis stricta, Creeping Wood Sorrel, **33**
Oxalis violacea, Violet Wood Sorrel, **18**
Oxalis, Yellow, Wood Sorrel, *Oxalis grandis,* **28**

Pale Jewelweed, Pale Touch-me-not, *Impatiens pallida,* **78**
Pale or White Violet, *Viola striata,* **17**
Pale Spiked Lobelia, *Lobelia spicata,* **49**
Pale Touch-me-not, Pale Jewelweed, *Impatiens pallida,* **78**
Panax quinquefolium, Ginseng, **32**
Panax trifolium, Dwarf Ginseng, **17**
Panicled Hawkweed, *Hieracium paniculatum,* **31**
Pansy, Field, *Viola Kitaibeliana,* **10**
Parthenium integrifolium, Wild Quinine, **59**
Partridge Pea, *Cassia fasciculata,* **61**
Pasture Rose, *Rosa carolina,* **54**
Pedicularis canadensis, Wood-betony or Lousewort, **20**
Pennywort, *Obolaria virginica,* **12**
Penstemon calycosus, Beard-tongue, **39**
Pepper Root, **5**
Perennial Sweet Pea, Everlasting Pea, *Lathyrus latifolius,* **41**
Periwinkle, Myrtle, *Vinca minor,* **8**
Petunia, Wild, *Ruellia strepens,* **39**
Phacelia purshii, Miami Mist, **23**
Phlox bifida, Creeping Phlox, Sand or Cleft Phlox, **9**
Phlox divaricata, Blue Phlox, Wild Sweet William, **16**
Phlox glaberrima, Smooth Phlox, **35**
Phlox, Blue, Wild Sweet William, *Phlox divaricata,* **16**
Phlox, Cleft, Creeping or Sand Phlox, *Phlox bifida,* **9**
Phlox, Creeping, Sand or Cleft Phlox, *Phlox bifida,* **9**
Phlox, Sand, Creeping or Cleft Phlox, *Phlox bifida,* **9**
Phlox, Smooth, *Phlox glaberrima,* **35**
Physalis heterophylla, Ground Cherry, **55**
Physostegia virginiana, False Dragonhead, Obedient-plant, **73**
Pickerelweed, *Pontederia cordata,* **64**
Pimpernel, Scarlet, *Anagallis arvensis,* **44**
Pink, Indian, Pink-root, Star-bloom, Worm-grass, *Spigelia marilandica,* **37**
Pink Lady's-slipper, Moccasin Flower, *Cypripedium acaule,* 25, **30**
Pink-root, Star-bloom, Worm-grass, Indian Pink, *Spigelia marilandica,* **37**
Pink, Rose, Rose Gentian, Bitter-bloom, *Sabatia angularis,* **60**
Pink, Deptford, *Dianthus armeria,* **42**
Pipes, Indian, Corpse Plant, *Monotropa uniflora,* **66**
Pitcher Plant, Northern, *Sarracenia purpurea,* **30**
Plantain, Downy Rattlesnake, *Goodyera pubescens,* **74**
Plantain-leaved Pussytoes, Early Everlastings, *Antennaria plantaginifolia,* **11**
Platanthera peramoena, also *Habenaria peramoena,* Purple Fringeless Orchid **60**
Pleurisy Root, Butterfly Weed, *Asclepias tuberosa,* **62**
Podophyllum peltatum, Mayapple or Mandrake, **20**
Pogonia, Nodding, Three-Birds Orchid, *Triphora trianthophora,* **66**
Pointed-leaved Tick Trefoil, *Desmodium glutinosum,* **52**
Pokagon State Park, l, 20, 21, 29, 38, 76
Polemonium reptans, Greek Valerian, **10**
Polygala sanguinea, Field Milkwort, **60**
Polygonatum biflorum, Soloman's Seal, **18**
Polygonum persicaria, Lady's Thumb, Redleg, Smartweed, **70**
Pontederia cordata, Pickerelweed, **64**
Poppy, Celandine, Wood Poppy, *Stylophorum diphyllum,* **7**
Poppy, Wood, Celandine Poppy, *Stylophorum diphyllum,* **7**
Potato Creek State Park, 62, 73
Potato-vine, Wild, Purple-throated Morning Glory, *Ipomoea pandurata,* **56**
Potentilla recta, Rough-fruited Cinquefoil, Five Fingers, **55**
Potentilla simplex, Trailing or Common Cinquefoil, Five Fingers, **28**
Prairie Dock, *Silphium terebinthinaceum,* 48, **54**
Prairie or White False Indigo, *Baptisia leucantha,* **48**
Prairie or Gray Coneflower, *Ratibida pinnata,* **54**
Prairie Pointers, Shooting Star, Mosquito Bills, Birdbills, American Cowslip, *Dodecatheon meadia,* **11**
Prairie Rose, *Rosa setigera,* **43**
Prairie Shoestring, Leadwort, *Amorpha canescens,* **52**
Prairie Trillium, *Trillium recurvatum,* 13, **16**
Primrose, Sundrops, *Oenothera fruticosa,* **45**
Primrose, Common Evening, *Oenothera biennis,* **72**
Prunella vulgaris, Heal-all, Selfheal, **47**
Puccoon, Hairy, *Lithospermum croceum,* **34**
Puccoon, Hoary, *Lithospermum canescens,* **12**
Purple Cress, *Cardamine douglassii,* **9**
Purple Fringeless Orchid, *Plantanthera peramoena* (also *Habenaria peramoena*), **60**
Purple Gerardia, *Gerardia purpurea,* **76**
Purple Loosestrife, Lythrum, *Lythrum salicaria,* **67**
Purple Rocket, *Iodanthus pinnatifidus,* **28**
Purple-headed Sneezeweed, *Helenium nudiflorum,* **61,** 75
Purple-stemmed Aster, Swamp Aster, *Aster puniceus,* **78**
Purple-throated Morning Glory, Wild Potato-vine, *Ipomoea pandurata,* **56**
Pussytoes, Plantain-leaved, Early Everlastings, *Antennaria plantaginifolia,* **11**
Puttyroot Orchid, Adam and Eve, *Aplectrum hyemale,* **26**
Pycnanthemum tenuifolium, Narrow Leaved Mountain Mint, **49**

Quaker Ladies, Bluets, Innocence, *Houstonia caerulea,* **11**

Queen Anne's Lace, Wild Carrot, Bird's Nest, *Daucus carota*, **58**
Quinine, Wild, *Parthenium integrifolium*, **59**

Ragwort, Golden, Groundsel, Squaw-weed, *Senecio aureus*, 12, **21**
Ragwort, Round-leaved, *Senecio obovatus*, **12**
Ranunculus hispidus, Buttercup, **15**
Ranunculus septentrionalis, Swamp Buttercup, **19**
Raspberry, Black, Thimbleberry, *Rubus* species, **43**
Ratibida pinnata, Gray or Prairie Coneflower, **54**
Rattlesnake-Master, Button Snakeroot, *Eryngium yuccifolium*, **48**
Rattlesnake Plantain, Downy, *Goodyera pubescens*, **74**
Red Clover, *Trifolium pratense*, **65**
Redleg, Lady's Thumb, Smartweed, *Polygonum persicaria*, **70**
Red Trillium, Toadshade, *Trillium sessile*, **23**
Rhexia virginica, Meadow Beauty, Deergrass, **68**
Rock Cress, Smooth, *Arabis laevigata*, **10**
Rocket, Dame's, Sweet Rocket, Violet Rocket, *Hesperis matronalis*, **27**, 28
Rocket, Purple, *Iodanthus pinnatifidus*, **28**
Rocket, Sweet, Dame's Rocket, Violet Rocket, *Hesperis matronalis*, **27**, 28
Rocket, Violet, Dame's Rocket, Sweet Rocket, *Hesperis matronalis*, **27**, 28
Rocket, Yellow, Winter Cress, *Barbarea vulgaris*, **8**
Rosa carolina, Pasture Rose, **54**
Rosa multiflora, Multiflora Rose, **27**
Rosa setigera, Prairie Rose, **43**
Rose Gentian, Rose Pink, Bitter-bloom, *Sabatia angularis*, **60**
Rose, Multiflora, *Rosa multiflora*, **27**
Rose, Pasture, *Rosa carolina*, **54**
Rose Pink, Rose Gentian, Bitter-bloom, *Sabatia angularis*, **60**
Rose, Prairie, *Rosa setigera*, **43**
Rose-mallow, Crimson-eyed, *Hibiscus palustris forma peckii*, **51**, 64
Rose-mallow, Swamp, *Hibiscus palustris*, **64**
Rosinweed, *Silphium integrifolium*, **54**
Rough Blazing Star, *Liatris aspera*, **73**
Rough-fruited Cinquefoil, Five Fingers, *Potentilla recta*, **55**
Roundheaded Bush Clover, *Lespedeza capitata*, **71**
Round-leaved Ragwort, *Senecio obovatus*, **12**
Royal Catchfly, *Silene regia*, **59**
Rubus sp., Black Raspberry, Thimbleberry, 43; Blackberry, **27**, **58**
Rudbeckia hirta, Black-eyed Susan, **62**
Rue Anemone, *Anemonella thalictroides*, **4**
Rue, Meadow, *Thalictrum dioicum*, **11**, 14
Ruellia strepens, Wild Petunia, **39**

Sabatia angularis, Rose Pink, Rose Gentian, Bitter-bloom, **60**
Salsify, Meadow, Goatsbeard, *Tragopogon pratensis*, **55**
Salt-and-pepper, Harbinger of Spring, *Erigenia bulbosa*, **3**
Sambucus canadensis, Elderberry, **79**
Sand Phlox, Creeping or Cleft Phlox, *Phlox bifida*, **9**
Sanguinaria canadensis, Bloodroot, **5**
Saponaria officinalis, Bouncing Bet, Soapwort, **65**
Sarracenia purpurea, Northern Pitcher Plant, **30**
Saururus cernuus, Lizard's Tail, **51**
Scarlet Pimpernel, *Anagallis arvensis*, **44**
Scutellaria incana, Downy Skullcap, **50**
Sedum ternatum, Wild Stonecrop, **17**
Seedbox, *Ludwigia alternifolia*, **61**
Selfheal, Heal-all, *Prunella vulgaris*, **47**
Senecio aureus, Golden Ragwort, Groundsel, Squaw-weed, 12, **21**
Senecio obovatus, Round-leaved Ragwort, **12**
Shades State Park, 80
Shakamak State Park, 60, 61, 64
Shooting Star, Mosquito Bills, Birdbills, Prairie Pointers, American Cowslip, *Dodecatheon meadia*, **11**
Short-stemmed Iris, *Iris brevicaulis*, **31**
Showy Lady's-slipper, *Cypripedium reginae*, **38**
Showy Orchis, *Orchis spectabilis*, **25**
Showy Trillium, *Trillium grandiflorum*, **20**
Signatures, Doctrine of, 78
Silene noctiflora, Night-Flowering Catchfly, **36**
Silene regia, Royal Catchfly, **59**
Silene stellata, Starry Campion, **53**
Silene virginica, Fire Pink, Catchfly, **28**
Silphium integrifolium, Rosinweed, **54**
Silphium terebinthinaceum, Prairie Dock, 48, **54**
Sisyrinchium angustifolium, Blue-eyed Grass, **42**
Sisyrinchium montanum, Blue-eyed Grass, **12**
Skullcap, Downy, *Scutellaria incana*, **50**
Skunk Cabbage, *Symplocarpus foetidus*, **1**
Smartweed, Lady's Thumb, Redleg, *Polygonum persicaria*, **70**
Smilacina racemosa, False Solomon's Seal, 21, **24**
Smilacina stellata, Starry False Solomon's Seal, **21**
Smooth Phlox, *Phlox glaberrima*, **35**
Smooth Rock Cress, *Arabis laevigata*, **10**
Smooth Yellow Violet, *Viola pensylvanica*, **13**
Snakeroot, Button, Rattlesnake-Master, *Eryngium yuccifolium*, **48**
Snakeroot, White, White Mistflower, *Eupatorium rugosum*, **68**
Sneezeweed, *Helenium autumnale*, **75**
Sneezeweed, Purple-headed, *Helenium nudiflorum*, **61**, 75
Snow Trillium, Dwarf White Trillium, *Trillium nivale*, **3**
Soapwort, Bouncing Bet, *Saponaria officinalis*, **65**
Solidago gigantea, Late Goldenrod, **80**
Solidago juncea, Early Goldenrod, **61**
Solidago species, Goldenrod, **75**
Solomon's Seal, False, *Smilacina racemosa*, **24**
Solomon's Seal, *Polygonatum biflorum*, **18**, 24
Sonchus asper, Sow Thistle, **55**
Sorrel, Creeping Wood, *Oxalis stricta*, **33**
Sorrel, Violet Wood, *Oxalis violacea*, **18**
Sorrel, Wood, Yellow Oxalis, *Oxalis grandis*, **28**
Sow Thistle, *Sonchus asper*, **55**
Spanish Bayonet, *Yucca filamentosa*, **44**
Spanish Needles, Sticktights, Tickseed Sunflower, *Bidens aristosa*, **69**
Specularia perfoliata, Venus' Looking-glass, **35**
Speedwell, Corn, *Veronica arvensis*, **10**
Spiderwort, *Tradescantia virginiana*, **42**
Spigelia marilandica, Indian Pink, Pink-root, Star-bloom, Worm-grass, **37**
Spiked Lobelia, Pale, *Lobelia spicata*, **49**
Spiraea latifolia, Meadowsweet, **52**
Spiral Orchid, *Spiranthes vernalis*, **47**
Spiranthes cernua, Nodding Lady's Tresses, **76**
Spiranthes tuberosa, Little Lady's Tresses, **74**
Spiranthes vernalis, Spiral Orchid, **47**
Spiraea tomentosa, Steeplebush, Hardhack, **53**
Spotted Jewelweed, Touch-me-not, *Impatiens capensis*, **63**
Spotted Joe-Pye-Weed, *Eupatorium maculatum*, **76**
Spotted Knapweed, *Centaurea maculosa*, **44**
Spreading Dogbane, *Apocynum androsaemifolium*, **44**
Spring Beauty, *Claytonia virginica*, 4, **6**
Spring Cress, *Cardamine bulbosa*, **21**
Spring Mill State Park, 12, 14, 39, 71
Spring or Dwarf Larkspur, *Delphinium tricorne*, **13**
Spurge, Flowering, *Euphorbia corollata*, **49**

Spurred Violet, *Viola rostrata,* **20**
Squaw-weed, Golden Ragwort, Groundsel, *Senecio aureus,* 12, **21**
Squirrel Corn, *Dicentra canadensis,* **6**
St. John's-wort, *Hypericum perforatum,* **57**
Star Chickweed, *Stellaria pubera,* **16**
Star Grass, Colicroot, *Aletris farinosa,* **49**
Star of Bethlehem, *Ornithogalum umbellatum,* **8**
Star-bloom, Worm-grass, Indian Pink, Pink-root, *Spigelia marilandica,* **37**
Stargrass, *Hypoxis hirsuta,* **12**
Starry Campion, *Silene stellata,* **53**
Starry False Solomon's Seal, *Smilacina stellata,* **21**
State Forests: Martin, 50; Yellowwood, 67
State Parks: Brown County, 42; Chain O'Lakes, 51; Clifty Falls, 5; Harmonie, 10, 56; Indiana Dunes, 30, 48, 65; Lincoln, 58, 78; McCormick's Creek, 11, 13, 15, 24, 66; Mounds, 26; Ouabache, 57; Pokagon, 1, 20, 21, 29, 38, 76; Potato Creek, 62, 73; Shades, 80; Shakamak, 60, 61, 64; Spring Mill, 12, 14, 39, 71; Tippecanoe River, 19, 49, 52, 53, 75; Turkey Run, 23; Whitewater, 65
Steeplebush, Hardhack, *Spiraea tomentosa,* **53**
Stellaria pubera, Star Chickweed, **16**
Stenanthium gramineum, Featherbells, Featherfleece, **40**
Sticktights, Spanish Needles, Tickseed Sunflower, *Bidens aristosa,* **69**
Stinking Benjamin, Wake-robin, *Trillium erectum,* **22**
Stonecrop, Wild, *Sedum ternatum,* **17**
Strawberry, Wild, *Fragaria virginiana,* **19**
Stylophorum diphyllum, Wood Poppy, Celandine Poppy, **7**
Summer Bluet, Large Houstonia, *Houstonia purpurea,* **39**
Sundrops, *Oenothera fruticosa,* **45**
Sunflower Crownbeard, *Verbesina helianthoides,* **37**
Sunflower, Tickseed, *Bidens aristosa,* **69**
Sunflower, Woodland, *Helianthus divaricatus,* **63**
Swamp or Purple-stemmed Aster, *Aster puniceus,* **78**
Swamp Buttercup, *Ranunculus septentrionalis,* **19**
Swamp Milkweed, *Asclepias incarnata,* **64**
Swamp Rose-mallow, *Hibiscus palustris,* **64**
Sweet Cicely, *Osmorhiza claytoni,* **22**
Sweet Clover, Yellow, *Melilotus officinalis,* **46**
Sweet Pea, Perennial, Everlasting Pea, *Lathyrus latifolius,* **41**
Sweet Pea, Wild, Goat's Rue, *Tephrosia virginiana,* **34**
Sweet Rocket, Dame's Rocket, Violet Rocket, *Hesperis matronalis,* **27,** 28
Sweet William, Wild, Blue Phlox, *Phlox divaricata,* **16**
Symplocarpus foetidus, Skunk Cabbage, **1**
Synandra hispidula, Synandra, **24**
Synandra, *Synandra hispidula,* **24**

Tall Bellflower, *Campanula americana,* **65**
Tall Coreopsis, *Coreopsis tripteris,* **72**
Tall Ironweed, *Vernonia altissima,* **79**
Tall Thistle, *Cirsium altissimum,* **79**
Tamarack, Eastern Larch, *Larix laricina,* 30
Taraxacum officinale, Dandelion, **8**
Teasel, *Dipsacus sylvestris,* **44**
Tephrosia virginiana, Goat's Rue, Wild Sweet Pea, Devil's Shoestrings, **34**
Teucrium canadense, Germander, Wood-sage, 52
Thalictrum dioicum, Meadow Rue, **11,** 14
Thimbleberry, Black Raspberry, *Rubus* species, **43**
Thimbleweed, *Anemone virginiana,* **57**
Thistle, Canadian, *Cirsium arvense,* **49**
Thistle, Nodding or Musk, *Carduus nutans,* **41**
Thistle, Sow, *Sonchus asper,* **55**
Thistle, Tall, *Cirsium altissimum,* **79**
Thorn Apple, Jimsonweed, *Datura stramonium,* **79**
Throughwort, Common, Boneset, *Eupatorium perfoliatum,* **78**
Three-birds Orchid, Nodding Pogonia, *Triphora trianthophora,* **66**
Three-lobed Violet, *Viola triloba,* **14**
Tickseed Sunflower, *Bidens aristosa,* **69**
Tick Trefoil, Pointed-leaved, *Desmodium glutinosum,* **52**
Tippecanoe River State Park, 19, 49, 52, 53, 75
Toadflax, Butter-and-eggs, *Linaria vulgaris,* **46**
Toadshade, Red Trillium, *Trillium sessile,* **23**
Tobacco, Indian, *Lobelia inflata,* **68**
Toothwort, Cut-leaved, *Dentaria laciniata,* **5**
Touch-me-not, Spotted Jewelweed, *Impatiens capensis,* **63**
Touch-me-not, Pale, Pale Jewelweed, *Impatiens pallida,* 63, **78**
Tradescantia virginiana, Spiderwort, **42**
Tragopogon pratensis, Goatsbeard, Meadow Salsify, **55**
Trailing or Common Cinquefoil, Five Fingers, *Potentilla simplex,* **28**
Trailing Arbutus, Mayflower, *Epigaea repens,* **2**
Trefoil, Birdfoot, *Lotus corniculatus,* **36**
Trifolium pratense, Red Clover, **65**
Trifolium repens, White Clover, **33**
Trillium cernuum, Nodding Trillium, 13
Trillium erectum, Wake-robin, Stinking Benjamin, **22**
Trillium flexipes, Drooping or Bent Trillium, **13**
Trillium grandiflorum, Showy Trillium, **20**
Trillium nivale, Snow or Dwarf White Trillium, **3**
Trillium recurvatum, Prairie Trillium, 13, **16;** *var. luteum,* Yellow Trillium, **23**
Trillium sessile, Toadshade, Red Trillium, **23**
Trillium, Drooping or Bent, *Trillium flexipes,* **13**
Trillium, Nodding, *Trillium cernuum,* 13
Trillium, Prairie, *Trillium recurvatum,* **16**
Trillium, Red, Toadshade, *Trillium sessile,* **23**
Trillium, Showy, *Trillium grandiflorum,* **20**
Trillium, Snow or Dwarf White, *Trillium nivale,* **3**
Trillium, Yellow, *Trillium recurvatum var. luteum,* **23**
Triosteum aurantiacum, Horse Gentian, Wild Coffee, **26**
Triphora trianthophora, Nodding Pogonia, Three-birds Orchid, **66**
Trout Lily, Fawn Lily, Adders-tongue, Dog-tooth Violet, *Erythronium americanum,* **6**
Trumpet Vine, *Campsis radicans,* **57**
Turkey Run State Park, 23
Turk's-cap, *Lilium superbum,* 48
Turtlehead, *Chelone glabra,* **77**
Twayblade, Large or Lily-leaved, *Liparis lilifolia,* **18,** 25
Twinleaf, *Jeffersonia diphylla,* **5**
Two-flowered Cynthia, Dwarf Dandelion, *Krigia biflora,* **18**

Umbrellawort, Heart-leaved, Wild Four-o'clock, *Mirabilis nyctaginea,* **34**
Urnula craterirum, Cup Fungus, 6
Uvularia perfoliata, Bellwort, **15**

Vaccinium macrocarpon, Large Cranberry, **33**
Valerian, Greek, *Polemonium reptans,* **10**
Valerian, *Valeriana pauciflora,* **24**
Valeriana pauciflora, Valerian, **24**
Venus' Looking-glass, *Specularia perfoliata,* **35**
Verbascum blattaria, Moth Mullein, **55**
Verbascum thapsus, Common Mullein, **57**
Verbena hastata, Blue Vervain, **53,** 58
Verbena stricta, Hoary Vervain, **58**
Verbesina helianthoides, Sunflower Crownbeard, **37**
Vernonia altissima, Tall Ironweed, **79**
Vernonia missurica, Missouri Ironweed, **75**
Veronica arvensis, Corn Speedwell, **10**
Veronicastrum virginicum, Culver's Root, **59**
Vervain, Blue, *Verbena hastata,* **53**

Vervain, Hoary, *Verbena stricta,* **58**
Vetch, Cow or Blue, *Vicia cracca,* **41**
Viburnum acerifolium, Maple-leaved Viburnum, Dockmackie, **28**
Viburnum, Maple-leaved, Dockmackie, *Viburnum acerifolium,* **28**
Vicia cracca, Cow Vetch, Blue Vetch, **41**
Vinca minor, Myrtle, Periwinkle, **8**
Viola conspersa, Dog Violet, **21**
Viola fimbriatula, Northern Downy Violet, **19**
Viola kitaibeliana, Field Pansy, **10**
Viola lanceolata, Lance-leaved or Water Violet, **33**
Viola missouriensis, Missouri Violet, **7**
Viola papilionacea, Common Blue Violet, Meadow Violet, **14, 16;** *var. priceana,* Confederate Violet, **9**
Viola pedata, Birdfoot Violet, **12**
Viola pensylvanica, Smooth Yellow Violet, **13**
Viola pubescens, Downy Yellow Violet, **20**
Viola rostrata, Spurred Violet, **20**
Viola sororia, Downy or Woolly Blue Violet, **6**
Viola striata, White or Pale Violet, **17**
Viola triloba, Three-lobed Violet, **14**
Violet Rocket, Dame's Rocket, Sweet Rocket, *Hesperis matronalis,* **27,** 28
Violet Wood Sorrel, *Oxalis violacea,* **18**
Violet, Birdfoot, *Viola pedata,* **12**
Violet, Common Blue, Meadow Violet, *Viola papilionacea,* **14, 16**
Violet, Confederate, *Viola papilionacea var. princeana,* **9**
Violet, Dog, *Viola conspersa,* **21**
Violet, Downy or Woolly Blue, *Viola sororia,* **6**
Violet, Downy Yellow, *Viola pubescens,* **20**
Violet, Green, *Hybanthus concolor,* **22**
Violet, Lance-leaved or Water, *Viola lanceolata,* **33**
Violet, Meadow or Common Blue, *Viola papilionacea,* **14, 16**
Violet, Missouri, *Viola missouriensis,* **7**
Violet, Northern Downy, *Viola fimbriatula,* **19**
Violet, Smooth Yellow, *Viola pensylvanica,* **13**
Violet, Spurred, *Viola rostrata,* **20**
Violet, Three-lobed, *Viola triloba,* **14**
Violet, Water or Lance-leaved, *Viola lanceolata,* **33**
Violet, White or Pale, *Viola striata,* **17**
Virginia Bluebells, Virginia Cowslip, *Mertensia virginica,* **16**
Virginia Waterleaf, *Hydrophyllum virginianum,* **13**

Wake-robin, Stinking Benjamin, *Trillium erectum,* **22**
Wand-like Bush Clover, *Lespedeza intermedia,* **72**
Water Lily, White, *Nymphaea odorata,* **51**
Water or Lance-leaved Violet, *Viola lanceolata,* **33**
Waterleaf, Appendaged, *Hydrophyllum appendiculatum,* **24**
Waterleaf, Broad-leaved, *Hydrophyllum canadense,* **40**
Waterleaf, Virginia, *Hydrophyllum virginianum,* **13**
White Avens, *Geum canadense,* **50**
White Baneberry, *Actaea pachypoda,* **15**
White Clover, *Trifolium repens,* **33**
White Fall Aster, *Aster pilosus,* **80**
White or Prairie False Indigo, *Baptisia leucantha,* **48**
White Lady's-slipper, *Cypripedium candidum,* **29**
White Mistflower, White Snakeroot, *Eupatorium rugosum,* **68**
White Monarda, Basil Balm, *Monarda clinopodia,* **40**
White or Pale Violet, *Viola striata,* **17**
White Snakeroot, White Mistflower, *Eupatorium rugosum,* **68**
White Water Lily, *Nymphaea odorata,* **51**
White Yarrow, *Achillea millefolium,* **43**
Whitewater State Park, 65
Wild Bergamot or Monarda, *Monarda fistulosa,* **53**
Wild Carrot, Queen Anne's Lace, Bird's Nest, *Daucus carota,* **58**
Wild Coffee, Horse Gentian, *Triosteum aurantiacum,* **26**
Wild Columbine, *Aquilegia canadensis,* **10**
Wild Four-o'clock, Heart-leaved Umbrellawort, *Mirabilis nyctaginea,* **34**
Wild Geranium, *Geranium maculatum,* **16**
Wild Ginger, *Asarum canadense,* **11**
Wild Hyacinth, *Camassia scilloides,* **23**
Wild Lettuce, Yellow, *Lactuca canadensis,* **46**
Wild Lily of the Valley, Canada Mayflower, *Maianthemum canadense,* **32**
Wild Lupine, *Lupinus perennis,* **35**
Wild Onion, *Allium stellatum,* **47**
Wild Petunia, *Ruellia strepens,* **39**
Wild Potato-vine, Purple-throated Morning Glory, *Ipomoea pandurata,* **56**
Wild Quinine, *Parthenium integrifolium,* **59**
Wild Senna, *Cassia hepecarpa,* **63**
Wild Stonecrop, *Sedum ternatum,* **17**
Wild Strawberry, *Fragaria virginiana,* **19**
Wild Sweet Pea, Goat's Rue, *Tephrosia virginiana,* **34**
Wild Sweet William, Blue Phlox, *Phlox divaricata,* **16**
Wingstem, *Actinomeris alternifolia,* **63**
Winter Cress, Yellow Rocket, *Barbarea vulgaris,* **8**
Wood Anemone, *Anemone quinquefolia,* **22**
Wood Poppy, Celandine Poppy, *Stylophorum diphyllum,* **7**
Wood Sorrel, Yellow Oxalis, *Oxalis grandis,* **28**
Wood Sorrel, Violet, *Oxalis violacea,* **18**
Wood-betony, Lousewort, *Pedicularis canadensis,* **20**
Wood-mint, Downy, *Blephilia ciliata,* **39**
Wood-sage or Germander, *Teucrium canadense,* **52**
Woodland Sunflower, *Helianthus divaricatus,* **63**
Woolly or Downy Blue Violet, *Viola sororia,* **6**
Worm-grass, Indian Pink, Pink-root, Star-bloom, *Spigelia marilandica,* **37**
Wormseed Mustard, *Erysimum cheiranthoides,* **19**

Yarrow, White, *Achillea millefolium,* **43**
Yellow Lady's-slipper, *Cypridium calceolus, var. parviflorum,* **29;** *var. pubescens,* **25**
Yellow Oxalis or Wood Sorrel, *Oxalis grandis,* **28**
Yellow Rocket, Winter Cress, *Barbarea vulgaris,* **8**
Yellow Sweet Clover, *Melilotus officinalis,* **46**
Yellow Trillium, *Trillium recurvatum var. luteum,* **23**
Yellow Wild Lettuce, *Lactuca canadensis,* **46**
Yellowroot, Goldenseal, *Hydrastis canadensis:* flower, **20;** fruit, **50**
Yellowwood State Forest, 67
Yucca filamentosa, Spanish Bayonet, **44**

Zizia aurea, Golden Alexanders, **8**

EDITOR: ROBERTA L. DIEHL
BOOK AND JACKET DESIGNER: SHARON L. SKLAR
PRODUCTION COORDINATOR: HARRIET CURRY
TYPEFACE: BEMBO
COMPOSITOR: G & S TYPESETTERS, INC.
PRINTER: TOPPAN